WOOD COMBUSTION
Principles, Processes, and Economics

WOOD COMBUSTION
Principles, Processes, and Economics

David A. Tillman
Amadeo J. Rossi
William D. Kitto

Envirosphere Company
Division of Ebasco Services, Incorporated
Bellevue, Washington

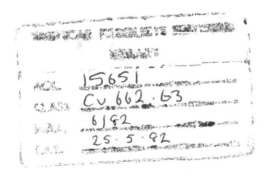

ACADEMIC PRESS   1981
A Subsidiary of Harcourt Brace Jovanovich, Publishers
New York London Toronto Sydney San Francisco

ACADEMIC PRESS, INC.
111 Fifth Avenue, New York, New York 10003

United Kingdom Edition published by
ACADEMIC PRESS, INC. (LONDON) LTD.
24/28 Oval Road, London NW1 7DX

Library of Congress Cataloging in Publication Data

Tillman, David A.
  Wood combustion.

  Includes index.
  1. Wood as fuel.   I. Rossi, Amadeo J.   II. Kitto,
William D.   III. Title.
TP324.T55   662'.65   81-10907
ISBN 0-12-691240-8   AACR2

PRINTED IN THE UNITED STATES OF AMERICA

81 82 83 84   9 8 7 6 5 4 3 2 1

For Darek

65
14

CONTENTS

PREFACE

Since the mid 1970s, wood fuel utilization has received increased emphasis in industrial circles, governmental agencies, and the academic community. The forces driving this emphasis are numerous, including: (1) the renewability of wood as a fuel; (2) the political and economic shocks buffeting imported oil and nuclear power; and (3) a plethora of applied research which has addressed many of the technical problems associated with using wood fuels.

The increased emphasis on wood fuels has focused on such diverse areas as cogeneration, alcohol fuels production, and the growing of select species (e.g., American Sycamore) to be used specifically as industrial fuel. Studies have addressed such questions as the economics of fuel supply, policy questions associated with wood fuel use, and technical problems surrounding wood combustion and conversion to gaseous and liquid fuels. Very few studies have attempted to integrate technical, environmental, and economic considerations, however. Further, few studies bring these disciplines to bear on policy questions in a rigorous fashion.

This book examines questions of present consumption, wood fuel supply, wood combustion, and the uses of wood combustion. It utilizes specific tools from such diverse disciplines as microeconomics, organic chemistry, thermodynamics, and financial theory. It focuses on near-term industrial applications in wood fuel utilization: raising steam, generating electricity, and producing select products by available conversion methods.

This book is an outgrowth of our collective work at Envirosphere Company, a Division of Ebasco Services; our work at the College of Forest Resources, University of Washington, on the Technology Assessment of Wood Fuels for the U.S. Congress Office of Technology Assessment; and on individual projects. It is the product of our efforts as the Energy Systems Planning Group at Envirosphere Northwest.

Each individual brought specific tasks to the project. David Tillman developed specific tools for analyzing combustion and heat release, NO_x formation, and financial analysis. Amadeo Rossi brought a detailed knowledge of fuel characteristics and particulate emissions to the project. William Kitto brought regulatory and environmental analysis skills as well as a broad understanding of forestry and the forest products industry.

Portions of the work have been published elsewhere, however, a great deal of unpublished material is contained in this volume. The basic combustion model was published in Wood Science, the scientific journal of the Forest Products Research Society. It has been improved here. Numerous tools such as the NASA combustion model, and the use of marginal costs curves not frequently employed in the analysis of wood combustion projects also are introduced along with attention to fundamental wood science concepts and mechanisms for pollution formation.

Considerable debate was focused upon units of measurement. It was decided that SI units would be preferred, with English units in parentheses. Moisture contents are expressed on a "green" or as-received basis, rather than on an oven-dry basis. Some exceptions have occurred where appropriate. The primary exception is in the area of economics, where oven-dry tonnes (and O.D. Tons) are the standard.

Wood Combustion: Processes, Uses, and Economics gained considerable strength from the assistance of numerous individuals including: Joe Franco, Ron Schnorr, Steve Simmons, and Joe Silvey of Envirosphere and Ebasco; Kyoti Sarkanen, Gerard Schreuder, and Ramsey Smith, University of Washington; Robert Jamison and Gordon Villesvik of Weyerhaeuser Company; Larry Anderson and David Pershing at the University of Utah; Kalevi Leppa of Ekono, Inc., and Charles Norwood, U.S. Department of Energy. It also received much attention and care from typists Millie Tillman and Rita Williams. Their help was invaluable.

With this as an introduction, we are pleased to present this volume. It provides a perspective on wood fuel, the most prominent biomass energy resource.

David A. Tillman
Amadeo J. Rossi
William D. Kitto

Envirosphere Company
A Division of Ebasco Services, Incorporated
Bellevue, Washington

CHAPTER 1

OVERVIEW OF WOOD FUELS ANALYSIS

I. THE PRESENT CONTRIBUTION OF WOOD FUELS TO THE U.S. ECONOMY

Wood fuels supplied 1.7×10^{18} J (Joules) or approximately 1.6×10^{15} Btu (1.6 quads) of energy to the U.S. economy in 1976 (Tillman, 1978). This represented approximately 3 percent of U.S. energy consumption. Wood fuel consumption increased to over 1.8×10^{18} J (1.7 quads) in 1978 (Schreuder and Tillman, 1980). Such is indicative of a larger trend that has prevailed over the last decade, when wood fuel consumption increased at a rate of approximately 0.1×10^{18} J/yr, a growth largely attributed to the forest products industry. As Table I shows, the pulp and paper industry was consuming more than 1.1×10^{18} J/yr and the lumber and plywood industries were using 0.3×10^{8} J/yr by 1978.

The data in Table I show that of the wood fuels consumed (energy basis) in the U.S., approximately 80 percent are used by the forest products industry, approximately 15 percent are used in residential applications, and the remaining 5 percent are

TABLE I. WOOD FUEL CONSUMPTION, 1978[a]

| | Wood energy consumption | | |
| | | | |
Economic sector	$\times 10^{18}$ J	$\times 10^{15}$ Btu (quads)	Percent
Pulp and paper	1.1	1.0	68
Lumber, plywood, particleboard	0.3	0.3	13
Other industry	0.1	0.1	6
Residential	0.3	0.3	13
TOTAL	1.8	1.7	100

[a]From Schreuder and Tillman (1980).

used in a variety of non-forest products industrial settings. Thus, the use of wood fuels for energy is influenced by activity in the forest products sector and is a function of the demand for lumber, plywood, and pulp and paper. Because of its dependence on the forest products industry, the production and consumption of wood fuels is sensitive to general economic conditions as well as individual economic phenomena such as housing starts.

Wood fuels are also important in cogeneration. This is illustrated by the forest products industry, which is not only the leader in wood fuels consumption, but also a leader in cogeneration (Resource Planning Associates, 1977). Thus, a relatively large portion of wood fuels support cogeneration systems, as Tables II and III show. All such cogeneration systems are steam-topping cycles. In these systems high-pressure steam is passed through a back pressure or extraction turbine. Combustion turbines fueled by gas from wood gasifiers have been proposed, but not yet employed.

TABLE II. COGENERATION ACTIVITY BY INDUSTRY IN THE UNITED STATES, 1976

Industry	Energy in steam used in cogeneration		Electricity generated by cogeneration	
	(10^{15} J)	Percent	(10^9 kWh)	Percent
Pulp and paper	236	40.4	10.14	32.3
Chemicals	190	32.5	11.44	36.5
Steel	101	17.3	7.00	22.3
Food	27	4.6	1.16	3.7
Petroleum refining	15	2.6	0.94	3.0
Textiles	15	2.6	0.69	2.2
TOTAL	584	100.0	31.37	100.0

Outside the forest products industry some wood fuel is used in the furniture and fixtures industry, which also generates its own residues (Ultrasystems, Inc., 1978). The other principal nonresidential consumers of wood fuels are utilities. It is estimated that five utilities generated some 66×10^6 kWh in 1978, consuming approximately 1×10^{15} J of wood fuel. The majority of these wood-fueled plants are condensing power units. The role of utilities in cogeneration, however, cannot be overlooked. Because electricity production is necessarily a part of a cogeneration project, electric utilities often have a prominent role in such projects. The role can range from extensive involvement through joint ownership to ancillary participation such as occurs when a utility merely serves as the organization "wheeling" power from producer to consumer.

The importance of wood fuels and spent pulping liquor in cogeneration is illustrated by the fact that these constitute about 20 percent of the fuel used in cogeneration systems. This

TABLE III. STEAM AND ELECTRIC POWER PRODUCED BY COGENERATION AS A FUNCTION OF FUEL UTILIZED, 1976[a]

Fuel used	Gross quantity raised ($\times 10^{15}$ J)	Steam generated ($\times 10^{15}$ J)	Electricity ($\times 10^9$ kWh)
Organic waste[b]	360	182	9.90
Residential oil	282	204	8.81
Coal	136	97	4.16
Other	150	101	8.50

[a]From Resource Planning Associates (1977)

[b] Virtually all of this is wood or spent pulping liquor.

reveals the economic sector-specific contribution of this biomass form.

Wood fuels supply about half of the energy required by the
forest products industry, although exact percentages vary consi-
derably for individual plants. The Weyerhaeuser Co., for example, is rated by Kidder, Peabody and Co. as the most energy-
efficient forest products company; and it obtains about 67 percent of its energy from wood fuels (Hollie, 1979). Typically,
the sawmills and pulp mills of such larger firms are integrated
complexes, making optimum use of the logs entering the plant.
Figure 1-1 is a simplified nonquantified materials and energy
flow for such an integrated mill. As can be seen from this
figure, the forest products industry has unique opportunities
for using increasing quantities of wood fuel.

A. Technology Implications of the Present Contribution

Direct combustion systems are the only type of wood energy
conversion systems employed today, even though gasification and
liquefaction processes are possible. The primary reason why
gasification systems have not been widely employed to date is
that most forest products industry energy demands are for steam
and electricity--needs met by combustion. Direct heat requirements such as veneer drying can be met by suspension-fired systems (Levelton and Associates, Ltd., 1978). Energy requirements
for pulp mill lime kilns can also be met by suspension-fired
wood (Leppa, 1980). Moreover, the size of gasifiers is limited,
restricting their use to applications where small systems are
most desirable. Thus, the market for gasification systems is
largely limited to non-forest industry applications.

Pyrolysis and liquefaction are largely directed at meeting
fuel-to-fuel conversion requirements. They must compete with
coal conversion systems, and many factors such as scale limitations preclude effective economic competition (Tillman, 1980).

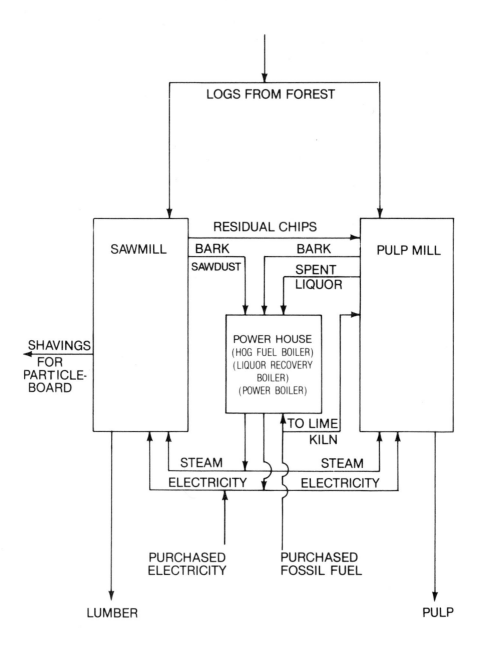

Thus, both liquefaction and gasification opportunities are limited, and combustion is expected to remain the technological backbone of the wood fuel utilization system.

B. Economic Implications of the Present Contribution of Wood Fuels

Increased deployment of combustion systems probably represents a change in demand for wood fuels, rather than a change in the quantity demanded. This shift is shown conceptually in Fig. 1-2. Evidence of this shift in demand abounds. It has been shown that, due to expanding markets, there are increasing sales of wood-fired boilers to non-forest industry firms (Tillman, 1978; Russell, 1978). It has also been shown that wood fuel is being used by the forest products industry as a direct substitute for purchased oil, natural gas, and electricity (Jamison et al., 1978). This substitution is occurring in other wood energy consuming sectors of the economy, and it is occurring despite rising wood fuel prices. For example, Burlington Electric (Vermont) is now paying approximately $1.51/J x 10^9 ($1.61/Btu x 10^6) for its wood fuel. The price for the fuel is rising at the same rate as coal in New England, yet increasing tonnages of wood are being consumed (U.S. Department of Energy, 1979). This change in demand suggests that wood fuels will be used increasingly in the coming decades.

FIG. 1-1. The integrated sawmill/pulpmill complex showing the flow of logs and residuals between components of the mill. Note the use of residual chips in pulping, and the flow of both bark and sawdust from the sawmill and bark and spent liquor from the pulpmill to the power house. The sawmill and pulpmill are the essential ingredients in a multiple-product integrated mill for the forest products industry.

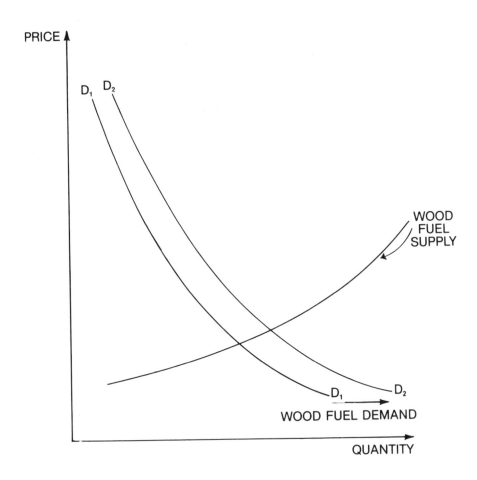

FIG. 1-2. Change in demand for wood fuel. Data exist to show that the demand for wood fuel has shifted to the right. Despite an increase in price for such fuel, more wood is being consumed for energy purposes now than previously.

II. THE FUTURE CONTRIBUTION OF WOOD FUELS

The future contribution of wood fuels to the U.S. economy has been the subject of much speculation. Tillman (1978), for example, made a conservative estimate of 4.3 x 10^{18} J in the year 2000. Bethel et al. (1979) estimated between 5 x 10^{18} and 10 x 10^{18} J may be consumed at the turn of the century, depending upon scenario. Others making comparable estimates include the National Academy of Sciences (1979) and Hayes (1979). These estimates reflect a growing realism compared to earlier, more optimistic estimates (which include forecasts of 20 x 10^{18} J/yr and higher).[*]

These lower projections result, in part, from reduced expectations of total energy consumption. Previous forecasts of 150 x 10^{18} to 200 x 10^{18} J/yr total energy consumption in the year 2000 are no longer considered realistic. Typical current estimates are in the range 75 x 10^{18} to 90 x 10^{18} J/yr. Thus, each fuel in the total energy mix must accept a reduced burden. Despite the reduced consumption forecasts, wood will probably increase its energy contribution both within and outside the forest products industry.

A. Future Use of Wood Energy in the Forest Products Industry

Wood energy use will increase most dramatically in the forest products industry for two reasons: (1) wood-based products are less energy intensive to produce and to use than mineral-based substitutes (Boyd et al., 1976), and (2) there is ample room for wood to increase its energy contribution within the forest products industry (Bethel et al., 1979).   Both factors

[*]See National Academy of Sciences (1979) for a complete discussion of forecasts.

TABLE   IV.   THE   ENERGY   INTENSITY   OF   SEVERAL   WOOD
AND COMPETING PRODUCTS[a]

Product (size)	Base	Energy cost ($\times 10^6$ J)
Exterior wall system (100 ft^2)	wood	2000-2700
Exterior wall system (100 ft^2)	concrete	17,100
Exterior wall system (100 ft^2)	metal	4800-5500
Milk container (1/2 gal)	wood (paper)	7.5
	plastic	11.3
Bag		
	wood (paper)	2.4
	plastic	3.7

[a]From Boyd et al. (1976) and Bethel et al. (1976).

will become increasingly important as the forest industry grows
and energy costs rise in the years ahead.

Studies have been conducted on the energy intensity of competing materials such as wood, steel, concrete, paper, and plas-
tics. Boyd et al. (1976) have calculated the gross energy intensity of various wall systems, while Bethel et al. (1976) have
shown values for paper and plastic bags, milk cartons, and other
products. Selected values are shown in Table IV. From this
table it becomes clear that wood-based products require less
energy to produce than mineral-based competitors. Thus they en-
joy a considerable advantage. Values in Table IV include both
the energy content of a product and the energy required for its
extraction, transportation, and manufacture.

TABLE V. ENERGY COMSUMPTION PER TON OF BLEACH KRAFT PULP PROCESS

	Process	
Fuel/energy	Average practice[a]	Best present practice[b]
Self generated fuels (in GJ)		
Black liquor	15.3	15.2
Hog fuel		3.2
Purchased fossil fuel (in GJ)	21.6	11.0
Self generated electricity (kWh)	470	949
Purchased Electricity (kWh)	500	101

[a]Source: Gyftopoulos et. al. (1974).
[b]Source: Hurley (1978).

Table IV, however, does not show the growing energy cost of mineral-based products. The process of petroleum extraction, for example, once relied only on natural oil field pressure. Today, steam injection into the oil field must be employed in order to increase recovery. Systems under development include the use of surfactants, CO_2 injection, and fire flooding for further enhanced recovery. All of these steps represent increasing financial and energy expenditures per barrel of oil recovered. Similarly, as more and more steel comes from taconite iron ore, additional energy must be expended in ore extraction, concentration, and sintering into pellets. In the case of steel, increased use of scrap metal mitigates this energy trend. Because of such factors, as energy becomes more expensive, the energy intensiveness of products will become more economically significant. Such trends support projections for wood product growth, where wood products will be substituted for their mineral competitors (Bethel et al., 1976; U.S. Forest Service, 1973).

The second force increasing the use of wood fuels is the opportunity to substitute more wood energy for fossil energy in many new and existing mills. Boyd et al. (1976) have demonstrated that sawmills could be virtually fuel self-sufficient. Gyftopoulos and others have shown that pulp mills could move substantially toward fuel self-sufficiency, as shown in Table V. As fossil fuels become more expensive, the movement toward essentially wood-fueled mills is expected to accelerate.

Outside the forest products industry, the use of wood fuels is also expected to accelerate slightly (Tillman, 1978). Local fuel supply-consumption imbalances can support some mills, such as the Russell Co. textile mill in Alexander City, Georgia (Russell, 1978); utilities, such as Eugene Water and Electric Board (EWEB); and other industrial and commercial establishments. Where such imbalances exist, they must be evaluated in terms of escalating residential demand for wood fuels. Nevertheless, in the event that plant operations become threatened by highly unstable fossil energy supplies, and the cost competitiveness of fuels becomes a secondary issue, this growth could indeed be rapid.

III. IDENTIFICATION OF SIGNIFICANT ISSUES

Wood fuel consumption projections between 4×10^{18} and 10×10^{18} J in the year 2000 represent opportunities or targets. Whether such targets are met depends upon a number of interrelated technical environmental and financial factors. Some of these factors are directly related to fuel supply and others to fuel utilization. In addressing significant issues, it is most important to understand the underlying factors rather than to attempt to assess the probability of meeting forecasts.

In some regions, where sawmills predominate, hog fuel remains the marginal wood fuel; however, in most regions, forest
residues are now the marginal increment. The fact that the
supply-consumption relationship is close to the limit of mill
residue supply and is already forcing the use of some forestry
residues has clear implications for wood fuel utilization
programs.

A. Fuel Supply Issues

In many cases the underlying issue limiting the potential
viability of a specific project is fuel supply. For new projects that non-forest industries are considering, the long-term
availability of fuel must be assured so that the plant can be
financed with confidence.

B. Technical and Environmental Issues

Because the marginal fuel is either hogged mill waste or
more expensive forest residues, maximizing energy efficiency in
the use of wood fuel is essential. Achieving high operating
efficiencies requires a basic understanding of the physical and
chemical structure of wood, the processess available for its
production, the pathways of biomass oxidation, and the systems
available for wood fuel utilization.

By means of these considerations, efficiencies of fuel utilization in raising heat and steam and generating electricity
can be calculated. Such efficiencies directly affect the financial desirability of small process energy systems, cogeneration systems, and condensing power systems. By weighing such
considerations, the viability of wood systems can be assessed.

C. Economic and Financial Issues

Whereas technical issues focus on the products and efficiencies of combustion, economic and financial issues focus on costs and cost competitiveness. The specific costs involved are those of money, of equipment, and of wood fuels. The cost of money is influenced by the business, financial, and technological risk of the investment. The cost of fuel is a function of the local marginal cost (supply) curve for wood fuel and the demand for that energy source.

Given appropriate technical, economic, and financial conditions, wood fuel systems can be attractive investments. Such conditions are best identified by case studies in wood fuel investments. Based on these data, the desirability of wood fuels can be assessed through trade-off analysis.

D. Retrospects and Prospect

The issues identified above cut across numerous disciplines: wood physics, chemistry, mechanical engineering, microeconomics, and finance. They are interrelated, and all bear on the future use of wood fuels. Each could be the focus of a book; however, all merit treatment here.

In order to deal with all such subjects here, an integrated approach is taken. Financial analysis is used, ultimately, to integrate the issues of fuel supply, combustion characteristics and systems, and costs. Finance is used for integrative purposes, as it determines whether new boilers, turbogenerators, and kilns will be built. Without such new systems, wood fuel consumption will remain at the present level of 1.8×10^{18} to 2.0×10^{18} J. With such new systems, the opportunities to double or triple that contribution level will be realized.

REFERENCES

Bethel, J. S., et al. (1976). "The Potential of Lignocellulosic Material for the Production of Chemicals, Fuels, and Energy." National Academy of Science (Committee on Renewable Resources for Industrial Materials), Washington, D.C.

Bethel, J. S., et al. (1979). "Energy from Wood." A Report to the Office of Technology Assessment, Congress of the United States. College of Forest Resources, Univ. of Washington, Seattle, Washington.

Boyd, C. W., Koch, P., McKean, H. B., Morschauser, C. R., Preston, S. B., and Wangaard, F. F. (1976). Wood for structural and architectural purposes. Wood and Fiber 8.

Gyftopoulos, E. P., Lazaridis, L. J., and Widmer, T. F. (1974). "Potential Fuel Effectiveness in Industry." Ballinger Publ. Co., Cambridge, Massachusetts.

Hayes, E. T. (1979). Energy resources available to the United States, 1985-2000. Science 203:233-239.

Hollie, P. G. (1979). Slow growth for a giant: Weyerhaeuser loses sales lead, finds its export plans delayed. New York Times, March 27.

Hurley, P.J. (1978). Comparison of Mill Energy Balance Effects of Conventional, Hydropyrolysis, and Dry Pyrolysis Recovery Systems. Institute of Paper Chemistry, Appleton, Wis.

Jamison, R. L., Methven, N. E., and Shade, R. A. (1978). "Energy from Forest Biomass." National Association of Manufacturers, Washington, D.C.

Leppa, K. (1980). Advanced drying process is key to burning peat, wood byproducts, in "Energy Systems Guidebook," pp. 59-61. McGraw-Hill, New York.

Levelton, B. H., and Associates, Ltd. (1978). "An Evaluation of Wood Waste Energy Conversion Systems." Environment Canada, Vancouver, B.C.

National Academy of Sciences (1979). "Energy in Transition—1985-2010." National Research Council, NAS, Washington, D.C.

Resource Planning Associates (1977). The potential for cogeneration development, in "Cogeneration of Steam and Electric Power" (R. Noyes, ed.), pp. 95-183. Noyes Data, Park Ridge, New Jersey.

Russell, B. (1978). Wood-fired steam plant at Russell Corporation, in "Wood Energy" (M. L. Hiser, ed.), pp. 113-117. Ann Arbor Science, Ann Arbor, Michigan.

Schreuder, G. F., and Tillman, D. A. (1980). Wood fuels consumption methodology and 1978 results, in "Progress in Biomass Conversion," Vol. 2 (K. V. Sarkanen and D. A. Tillman, eds.), pp. 60-88. Academic Press, New York.

Tillman, D. A. (1978). "Wood as an Energy Resource." Academic Press, New York.

Tillman, D. A. (1980). Fuels from waste, in "Kirk-Othmer Encyclopedia of Chemical Technology," Vol. 2, 3rd ed. Wiley, New York.

Ultrasystems, Inc. (1978). "Wood Energy for Small Scale Power Production in North Carolina." North Carolina Dept. of Commerce, Raleigh.

U.S. Department of Energy (1979). "Cost and Quality of Fuels for Electric Utility Plants—August 1979." USGPO, Washington, D.C.

U.S. Forest Service (1973). "The Outlook for Timber in the United States." U.S. Dept, Washington, D.C.

CHAPTER 2

PROPERTIES OF WOOD FUELS

I. INTRODUCTION

In order to understand the utility and behavior of wood as an energy source, it is essential to define those anatomical, physical, and chemical properties of wood which influence its behavior as a fuel. Such properties determine its materials handling, combustion, and heat release characteristics. Anatomical characteristics of importance include the structure of wood fibers and the pathways for moisture movement. Physical properties of interest include moisture content, specific gravity, void volume, and thermal properties. Chemical characteristics of importance include the summative analysis (holocellulose and lignin content), proximate and ultimate analysis, and higher heating value.

Trees belong to the vegetable kingdom, whose major classifications are Thallophytes, Bryophytes, Pteridophytes, and Spermaphytes. The Thallophytes are characterized by algae, fungi, and bacteria; the Bryophytes consist of mosses; and the Pteriodphytes comprise ferns. The Spermaphytes include all seed-

bearing plants and are further subdivided into Gymnosperms and
Angiosperms. The Gymnosperms comprise the coniferous or softwood
species of trees, and the Angiosperms the deciduous or hardwood
species.

Softwoods and hardwoods are both anisotropic (properties dependent on direction) and hygroscopic (loses or gains moisture)
in nature. Their chemical components include cellulose, hemicelluloses, and lignin in varying amounts depending on species.
In general, hardwoods contain more holocellulose (i.e., carbohydrates) and less lignin than softwoods.

The anatomical, physical, and chemical properties of wood
need to be understood before any thorough treatment of materials
handling and combustion theory can occur. Anatomical properties
identify the macroscopic structures of softwoods and hardwoods,
physical properties deal with the relationships between specific
gravity and moisture content, and chemical properties refer to
carbohydrate and lignin structures and contents as related to
fuel reactivity and heating value. Although the following discussion is not exhaustive, it will provide the reader with a basic understanding of wood and its fuel-related properties.

II. ANATOMICAL CHARACTERISTICS OF WOOD FUEL

Anatomical characteristics of wood fuel directly affect the
structure of wood and determine many of the wood fuel-moisture
relationships. They are most readily understood in terms of
Fig. 2-1. Figure 2-1a presents a traverse or cross-sectional
view of a tree stem (trunk or bole). The outside layer of a
tree is covered with bark, the outermost layer of which is made
up of a corky material consisting of dead tissue, while the
inner bark (phloem) is composed of soft living tissue. Between
the bark layer and the stemwood (xylem) interface is a thin in-
distinguishable layer, the cambium. The cambium is living tis-

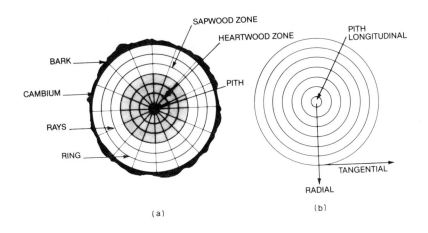

FIG. 2-1. Characteristics of stem wood: (a) a transverse cross section of wood fuel, including the location of significant items such as bark, heartwood, sapwood, and pith; (b) the direction associated with wood fuel analysis.

sue that generates new phloem and xylem tissue needed for tree growth. Located between the cambium and the pith is the main stemwood or xylem. The stemwood is composed of two distinct zones: sapwood and heartwood. The sapwood is composed of functioning dead tissue. It is responsible for moisture and nutrient movement and has a high moisture content. The heartwood extends from the pith to the sapwood zone and is composed of dead tissue. It is characterized by a low moisture content and high extractive content. Located at the center of the tree is the pith. Next to the pith is juvenile wood, which is characterized by low specific gravity and cellulose content. Juvenile wood is produced early in a tree's life and with each successive growth ring. Growth rings are also shown in Fig. 2-1a. Each year the tree adds one ring, which corresponds to that year's growth. Wood rays originate from the cambium and extend to the pith and bark, running perpendicular to the growth rings. The rays provide a means of food and nutrient storage and movement throughout the tree. As sapwood changes into heartwood, the wood rays die and become plugged with extractives (Panshin and DeZeeow, 1970).

Figure 2-1b defines the directions of a tree with respect to its cross section. The direction tangent to the growth rings is known as the tangential direction, with the radial direction being perpendicular to it and parallel to the wood rays. The longitudinal direction is perpendicular to the cross section and parallel to the pith.

A tree grows from the cambium layer by producing xylem cells, which grow inward, and the bark cells, which grow outward with respect to the cambium. Each year a layer of wood grows over the preceding year's growth, adding a growth ring as portrayed in Fig. 2-2. As the diameter of the stem increases, the bark is pushed outward, with some falling off and other taking on a pattern characteristic to each species of tree. As the tree gets older, the lower branches fall off. With successive layers of growth, the branch stubs are covered with new wood.

A. Structure of Softwoods

The principal building blocks of coniferous woods (soft-woods) are the tracheids, wood rays, and resin canals, with a volumetric composition for white pine of 93, 6, and 1 percent, respectively (Panshin and DeZeeow, 1970). A typical cross-sectional view of a softwood species is given in Fig. 3a. The tracheids are oriented longitudinally and are divided into two distinct zones: the earlywood (springwood) and latewood (summerwood). The transition between the springwood and summerwood zones can be abrupt as in Fig. 2-3a or gradual, depending on the growing stresses involved.

Summerwood is also characterized by an increase in cell wall thickness, as shown in Fig. 2-3a. Tracheids range in length from 7 mm (0.28 in.) for redwood to 3.5 mm (0.14 in.) for white spruce, with a tangential diameter ranging from 80 μm for redwood to 35 μm for white spruce (Wenzl, 1970; Panshin and

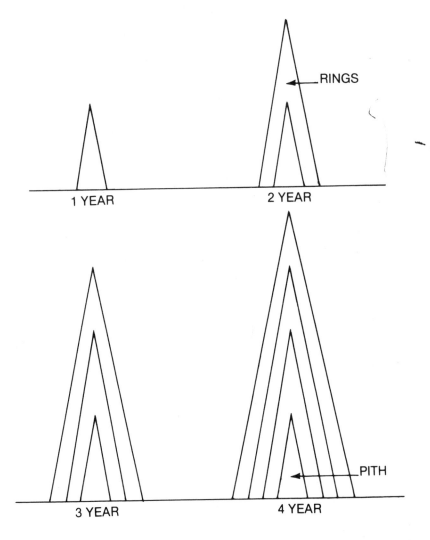

FIG. 2-2. Idealized tree growth, showing how rings are added to wood fuel.

DeZeeow, 1970). This range of values is typical for most commercial softwood species. The main function of the tracheids, in a tree's early development, is the conduction of water through their lumina. The conductance of water between adjacent tracheids takes place through pits located along the length and at the tips of the tracheid, as shown in Fig. 2-3c. As the tree ages, the tracheids' main function is to add structural support to the tree.

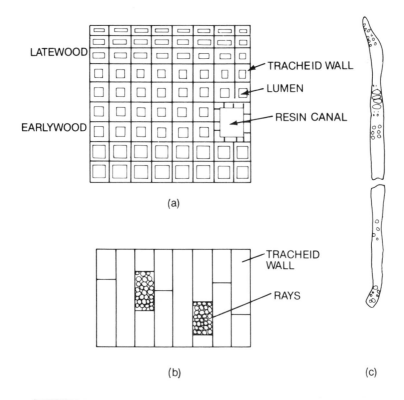

(a)

(b)   (c)

FIG. 2-3. Representative structure of softwoods. The concepts illustrated are essential to understanding the movement of moisture within wood fuel. (a) Transverse cross section; (b) tangential plane; (c) tracheid radial aspect (from Panshin and DeZeeow, 1970).

Figure 2-3b shows the ray ends in the tangential direction. The rays are composed of parenchyma cells and ray tracheids. The parenchyma cells are connected to the tracheids by large window-type pits and contain living matter essential for tree growth. The ray tracheids are connected to the tracheids by small bordered pits. Resin canals are essentially like the wood rays except they are oriented in the longitudinal direction, as depicted in Fig. 2-3a. The tracheids, wood rays, and resin canals form the major building blocks for softwoods, tying the tree together into a complete network via their pits (Cowling and Kirk, 1976).

B. Structure of Hardwoods

Hardwoods are more complex than softwoods, with vessel elements, fiber tracheids, longitudinal parenchyma [and wood rays, with a typical volumetric composition for sweet gum of 54.9, 26.3, 0.5, and 18.3 percent, respectively (Panshin and DeZeeow, 1970)]. A cross section of a typical hardwood species is shown in Fig. 2-4a, while a representation of a vessel element and fiber tracheid is presented in Fig. 2-4b. The vessels run parallel to the tree axis and are long continuous tubes composed of many vessel elements, which serve to conduct water. The length of the vessel elements range from 0.18 mm (0.007 in.) for black locust to 1.33 mm (0.05 in.) for black topelo.

In the heartwood of hardwoods, the vessel elements are usually closed, being plugged by a gumlike substance. The fiber tracheids are thick walled with a small lumen, typical fiber lengths range from 00.9 mm (0.035 in.) to 2.0 mm (0.078 in.). The rays, although more complex, provide the same function through their pits as softwoods.

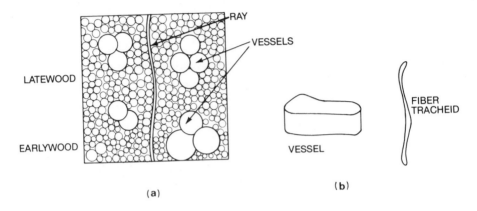

FIG. 2-4. Representative structure of hardwoods. (a) Transverse cross section; (b) vessels and wood fibers (from Panshin and DeZeeow, 1970).

C. Fuel Implications

The fuel implications of anatomical structure include the fact that wood can both adsorb and absorb moisture into the traechids and lumina. Once moisture exists in those structures, its removal is a difficult and energy-intensive process. Beyond moisture sorption in the living tree, the tracheids, lumen, and rays define the pathways of moisture migration when fuel is sub- jected to combustion conditions. They inhibit rapid evaporation of the moisture in large fuel particles and alter the heat transfer properties of the fuel particle itself. These factors are explored more thoroughly next.

III. PHYSICAL PROPERTIES OF WOOD FUEL

This section defines and discusses the relationship between moisture content, and specific gravity as related to void volume, and the thermal energy needed to evaporate water from wood. This is particularly important in the combustion of wet wood, since moisture content directly affects the net heating value of the wood fuel, the pathways of combustion, and the production of useful heat (efficiency of combustion).

A. Moisture Content

Wood is hygroscopic in nature and absorbs and adsorbs water. Water is absorbed by capillary uptake in the lumen of the wood structure; this absorbed water is known as free water. The sorption of water into the wood microstructure is adsorption, and adsorbed water is called "bound water." Adsorption is characterized by the release of heat due to the formation of hydrogen bonds and the interaction of Van der Waals forces. As adsorption takes place, the wood material swells. When the force for forming hydrogen bonds is balanced by the resulting force of further expansion, the fiber saturation point (FSP) is reached. The FSP for wood ranges from 23 to 25 percent moisture content green basis (defined below) with water being adsorbed below the FSP and absorbed above the FSP.

Moisture content of wood is typically reported as percentage moisture content oven dry and percentage moisture content green:

$$MC_{OD} = \text{weight}(H_2O)/\text{weight}(\text{dry wood matter}) \qquad (2-1)$$

$$MC_G = \text{weight}(H_2O)/[\text{weight}(\text{dry wood}) + \text{weight}(H_2O)] \qquad (2-2)$$

Moisture content green basis and dry basis can be interconverted using the equations below, which are derived from the above relationships:

$$MC_{OD} = 100MC_G/(1 - MC_G) \qquad (2\text{-}3)$$
$$MC_G = 100MC_{OD}/(MC_{OD} + 100) \qquad (2\text{-}4)$$

A practical limit of 400 percent MC_{OD} basis exists for several species, corresponding to a green moisture content of 80 percent. This is defined as a practical limit or as the maximum moisture content (MC_{MAX}).

TABLE I. MOISTURE CONTENT AND SPECIFIC GRAVITY OF SELECTED WOOD SPECIES[a]

Species	Moisture content, percent OD[b]		Specific gravity
	Heartwood	Sapwood	$S_{o,g}$
Softwoods			
cedar, incense	40 (23)	13 (68)	0.35
cedar, western	58 (37)	249 (71)	0.31
Douglas fir, coast	37 (27)	115 (53)	0.45
hemlock, western	85 (46)	170 (63)	0.42
pine, longleaf	31 (23)	106 (52)	0.54
redwood, old growth	86 (46)	210 (68)	0.38
Hardwoods			
ash, white	46 (31)	44 (30)	0.55
aspen	95 (49)	113 (53)	0.36
cottonwood	162 (62)	146 (59)	0.56
maple, silver	58 (37)	97 (49)	0.44
walnut, black	90 (47)	73 (42)	0.51

[a]From Skaar (1972) and USDA Forest Service (1974).
[b]Values in parentheses are moisture content green.

The MC_{MAX} of wood is related to its specific gravity S. The range in values is represented by black ironwood, where S = 1.18 and a MC_{OD} = 26 percent, and balsa wood, where S = 0.20 and MC_{OD} = 400 percent (Siav, 1971). Typical values for some commercial species are presented in Table I. The difference in moisture content between heartwood and sapwood is quite pronounced in the case of softwoods but is less significant in hardwoods. In the case of hardwoods, the MC of the heartwood can be higher than that of the sapwood, as is the case for cottonwood in Table I. The moisture content of the sapwood varies depending on region and site of the tree, while that of the heartwood is fairly uniform throughout the year due to the restricted movement of water. After a tree is harvested and chipped, its moisture content decreases. Wood protected from the elements will eventually reach an equilibrium moisture content (EMC).

The moisture content of wood fuels varies substantially among species and within the same tree. The moisture content is important since energy must be consumed to evaporate the water during fuel combustion. Therefore, it is wise either to store the fuel under cover in order to reach a lower moisture content whenever practical.

By way of comparison, the green moisture content of coal may be defined as its bed moisture content. Typical values are given in Table II with the moisture content increasing from anthracite to lignite due to the decrease in specific gravity. Coal gains very little water above its bed moisture content value. The moisture gain is restricted to the surface and thus is easily removed.

B. EVAPORATION OF WATER

The energy required to dry wood is a function of (1) the energy needed to heat the wood substance to 373°K (212°F); (2)

TABLE II. BED MOISTURE CONTENT OF SELECTED COALS[a]

Type	Moisture (percent)	Location (county, state)
Anthracite	2.5	Schuyllall, PA
Bituminous		
low volatile	1.3	Cambria, PA
medium volatile	1.5	Somerset, PA
high volatile	5.8	Williamson, IL
Subbituminous		
A	14	Musselshell, MT
B	25	Sheridan, WY
C	31	Campbell, WY
Lignite A	37	Mercer, ND

[a]From Babcock and Wilcox (1978).

the energy needed to heat the water to 373°K; (3) the heat of vaporization of water at 373°K; and (4) the energy necessary for desorption of the bound water in wood. In this discussion heat losses and thermal conductivity will not be addressed. In general, as the moisture content of wood increases, the thermal conductivity also increases, as discussed in Chapter 4. The energy discussed here is the energy needed to evaporate water from wood at 373 K.

The energy needed to heat the wood substance is a function of the specific heat of wood. Equation (2-5) is an expression for the specific heat of wood in a temperature range 273-373°K (32-212°F) (Dunlap, 1912). For the temperature (the variable T), it is sufficiently accurate to use an average of the final and initial temperatures of the wood:

$$C_{wood} = 0.0635 + 2.769 \times 10^{-4}(T - 273) \qquad (2-5)$$

where C_{wood} is the specific heat of wood (kJ/kg) and T is temperature (K).

The energy needed to heat the water to 373°K (212°F) is related to the specific heat of water, which ranges from 4.18 to 4.22 kJ°kg from 273 to 373°K (32-212°F). An average value of 4.2 kJ°kg°C can be used for calculating purposes.

The energy needed to evaporate water once it has reached 373°C is defined as H_{vap}, the heat of vaporization, which for water at 373 K is 2.26 MJ/kg.

Energy is needed to overcome H_w, the heat of wetting in wood, in order to free the bound water. Typically, for MC_g 4-12 percent, the energy needed is 35.8-14.3 kJ/kg, respectively; and for a green moisture content of 12-30 percent, the energy needed is 9.5-2.4 kJ/kg, respectively (adopted from Stamm, 1935).

The overall energy needed to evaporate water from wood then becomes a sum of (1) energy needed to heat wood, Q_1; (2) energy needed to heat water, Q_2; (3) heat of vaporization of water, Q_3; (4) energy to overcome the heat of wetting, Q_4. The respective equations are

$$Q_1 = 1 - (MC_g/100)(100°C - T_F)(C_{wood})(W_F) \qquad (2-6)$$
$$Q_2 = (MC_g/100)(100°C - T_F)(C_{water})(W_F) \qquad (2-7)$$
$$Q_3 = (MC_g/100)(H_{var})(W_F) \qquad (2-8)$$
$$Q_4 = (MC_g/100)(H_{wet})(W_F), \qquad MC_g \leq 23 \text{ percent} \qquad (2-9)$$

where MC_g is the green moisture content (percent); T_F the temperature of the fuel (°K); W_F the weight of the fuel, as is basic (kg); C_{wood} the specific heat of wood (kJ/kg°C); C_{water} the specific heat of water (kJ/kg°C); H_{vap} the heat of vaporization of water (kJ/kg H_2O); and H_{wet} the heat of wetting (kJ/kg H_2O).

The overall energy in a renumeration is given by

$$Q_{total} = Q_1 + Q_2 + Q_3 + Q_4 \qquad (2-10)$$

with Q_{total} the amount of energy necessary to evaporate water from wood assuming the initial condition described above.

As an example, assume 1000 kg of hog fuel at 45 percent MC_g and a temperature T = 293 K. We assume C_{wood} = 0.80 KJ/kg°C, H_{vap} = 2.26 MJ/kg, and H_{wet} = 15.5 as a weighted average.

Then substituting the given values into Eq. (2-10), one gets

$$Q_{total} = (3520 + 151,200 + 1,017,900 + 4650) \text{ kJ/1000 kg}$$
$$= 1.18 \text{ GJ/1000 kg fuel}$$
$$= 508 \text{ Btu/lb fuel}$$

Alternatively, the value can be expressed as 2.62 MJ/kg H_2O (1130 Btu/lb H_2O).

As can be seen, the major losses are in the heat of vaporization and specific heat of water in raising the temperature of water to 373°K.

C. Specific Gravity

Specific gravity provides a unitless measure for relating weight and volume. As such it provides a convenient approach to analyzing fuel storage and materials-handling requirements for the boiler. Further, it provides a basis for calculating maximum moisture content (MC_{MAX}) and fractional void volume (FVV). The importance of maximum moisture content has been shown previously. Fractional void volume determines the insulative properties of wood fuel, thus influencing the pathways of combustion.

Specific gravity (S) varies among different trees of the same species and also within the same tree. Typically, it increases from the pith to the bark, decreases with increasing height, and varies within the same growth ring. The specific gravity of solid cell wall substance is 1.54, while that of Douglas fir is 0.45, indicating that wood has a substantial void area with a large capacity to hold water and air (Kellog and Wangaard, 1969).

Specific gravity is defined as the unitless ratio

$$S = (W_w/V_w)W_{H_2O}/V_{H_2O} \qquad (2\text{-}11)$$

where W_w is the weight of the dry wood, V_w the volume of the dry wood, W_{H_2O} the weight of an equal volume of water, and V_{H_2O} the volume of water.

Since water has an approximate specific gravity of 1.0, Eq. (2-11) can be written

S = oven dry weight of wood/volume of water displaced   (2-12)

Multiplying the specific gravity by 100 kg/m^3 (62.4 lb/ft^3) gives the density of the wood sample.

Three typical ways of expressing the specific gravity of wood are listed below with the most common way being $S_{o,a}$ and $S_{o,g}$:

(1) $S_{o,o}$, specific gravity, oven dry weight and volume;
(2) $S_{o,a}$ is specific gravity, oven dry weight and air dry volume MC < FSP; and
(3) $S_{o,g}$ is specific gravity, oven dry weight, and green volume, MC > FSP.

Depending on the volume of the sample, and since wood swells with increasing moisture content (up to the FSP), the above distinction is necessary. Since $S_{o,g}$ is based on green volume (MC > FSP) and maximum swelling, the specific gravity is lower than that of $S_{o,a}$ (MC < FSP).

The following expression can be used to relate $S_{o,a}$ and $S_{o,g}$. Note that the density of the bound water (i.e., MC<FSP) is dependent on the moisture content (Skaar, 1972):

$$S_{o,g} = S_{o,a}/[1 - S_{o,a}(MC_a/100P_{BWa} - 0.2695]^{-1} \qquad (2\text{-}13)$$

where P_{BWa} is the density of bound water at MC_a from Fig. 2-5, and MC_a the moisture content oven dry basis (MC < FSP), percent.

Equation (2-13) is a simplified expression of a more general relation given in Eq. (2-14), which can be used to convert between any two bases (i.e., $S_{o,a}$ or $S_{o,b}$, where a and b are any two moisture content values less than the fiber saturation point):

$$S_{o,b} = S_{o,a} \times (1 - S_{o,a}[(MC_a/100P_{BWa})$$
$$-(MC_b/100P_{BWb})] \qquad (2-14)$$

where P_{BWa} is the density of bound water at MC_a, P_{BWb} the density of bound water at MC_b, MC_a the moisture content oven dry basis, percent, and MC_b the moisture content oven dry basis, percent, with $MC_a < MC_b$.

The specific gravity of the cell wall substance has been shown to be roughly 1.54. With this as the $S_{o,g}$ (or $S_{o,a}$) value of specific gravity, one calculates the maximum moisture content of wood on an oven dry basis in the following way (Skaar, 1972):

$$MC_{max} = (1/S_{o,g}) - (1/1.54) \times 100 \qquad (2-15)$$

D. Fractional Void Volume

The fractional void volume is the amount of air space in the wood after accounting for the volume of solid wood substance, the volume of bound water, and the volume of free water. The calculation is dependent on the specific gravity, the moisture content, and the density of bound water as expressed in (Skaar, 1972)

$$FVV = 1 - (S_{o,o}/1.54) + (S_{o,a}MC_a/100P_{BWa})$$
$$+[S_{o,g}(MC_g - 30)/100] \times 100 \qquad (2-16)$$

where FVV is fractional void volume, percent, and $S_{o,o}$ is specific gravity, oven dry weight and volume.

If the moisture content of the wood is below the fiber saturation point, then the third term in the equation drops out.

Also, $S_{o,o}$ can be calculated from Eq. (2-14) using the corresponding value for the density of bound water.

E. Physical Properties and Fuel Values

Physical properties largely determine the infuences of moisture on fuel heating value and the initial pathways of wood combustion. The moisture content, and whether the moisture is "free" or "bound" water, influences the energy losses associated with the moisture in wood combustion. Other physical properties, such as specific gravity, influence fuel storage and handling requirements, maximum moisture content, and fractional void volume as insulative properties. These also influence the pathways of wood combustion.

IV. CHEMICAL PROPERTIES OF WOOD FUEL

Combustion is a series of chemical reactions by which carbon is oxidized to carbon dioxide, and hydrogen is oxidized to water. The pathways of combustion are determined largely by the structure of any combustible molecule, particularly as it determines the location and accessibility of the carbon and hydrogen. Similarly, the higher heating value is determined by the relative proportions of carbon in a solid fuel (Tillman, 1978).

Because combustion is a series of chemical reactions, the chemical structure and composition of wood fuels merit careful consideration here. This analysis provides the foundation for an evaluation of combustion pathways, patterns of heat release, and formation of airborne emissions--topics considered in Chapters 4, 5, and 6.

The major chemical constituents of wood are cellulose, hemicelluloses, and lignin. Extractives are a minor constituent of most wood species and will not be discussed here.[*] Generally, softwoods have 40-45 percent cellulose, 24-37 percent hemicelluloses, and 25-30 percent lignin, with the hardwoods containing approximately 40-50 percent cellulose, and 22-40 percent hemicellulose (Shafizadeh and DeGroot, 1976). As can be seen, the hardwoods generally have a higher hemicellulose and a lower lignin content, and sometimes a slightly higher cellulose content. Table III presents a summative analysis of several wood species.

A. Holocellulose

The word holocellulose refers to the total carbohydrate of wood, which is represented by cellulose and the hemicellulose fractions. Hemicelluloses are further characterized by such constituents as xylan, galactan, and mannan.

Figure 2-5a represents the structure of the cellulose polymer. It is characterized by a repeating unit, a reducing and nonreducing end-group, and a β-1-4 glucosidic linkage. The actual linkage formed between adjacent anhydroglucose units, commonly called a glucoridic linkage, is generically equivalent to an ether linkage in stability. The repeating unit of cellulose is anhydroglucose in cellulose. However, it exists in the anhydro form. The number of repeating units or the degree of polymerization (DP) can range from 100 to 10,000 units. The nonreducing end-group is stable while the reducing end group is susceptible to acidic and basic attack. The cellulose polymer could be classified as a polyalcohol (Cowling and Kirk, 1976). The many hydroxyl (OH) functional groups form hydrogen bonds with water, as well as hydrogen bonds between adjacent anhydroglucose units and also adjacent cellulose chains. The OH groups

[*]In general, wood with a high extractive content has a higher heating value.

TABLE III. CHEMICAL COMPOSITION OF 10 SPECIES OF NORTH AMERICAN WOODS[a],[b]

	Cellulose (percent)	Lignin (percent)	Hemi-celluloses (percent)
Beech	45.2	22.1	32.7
White birch	44.5	18.9	36.6
Red maple	44.8	24.0	31.2
Eastern white cedar	48.9	30.7	20.4
Eastern hemlock	45.2	32.5	22.3
Jack pine	45.0	28.6	26.4
White spruce	48.5	27.1	21.4

[a]Extractive free basis, dry weight basis.
[b]From Wenzl (1970).

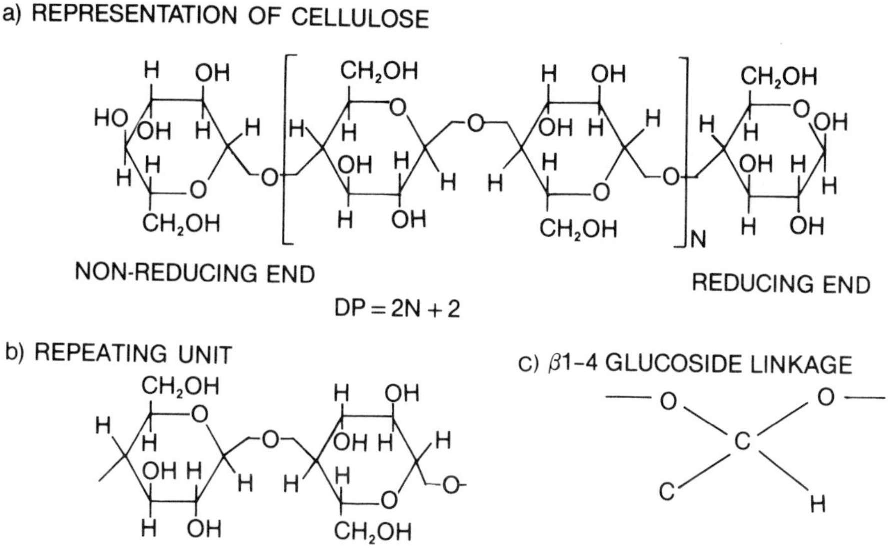

a) REPRESENTATION OF CELLULOSE

NON-REDUCING END

$DP = 2N + 2$

REDUCING END

b) REPEATING UNIT

c) β1-4 GLUCOSIDE LINKAGE

FIG. 2-5. The structure of cellulose. Note that the functional groups are OH and CH_2OH. Also note the placement of oxygen in ether linkages.

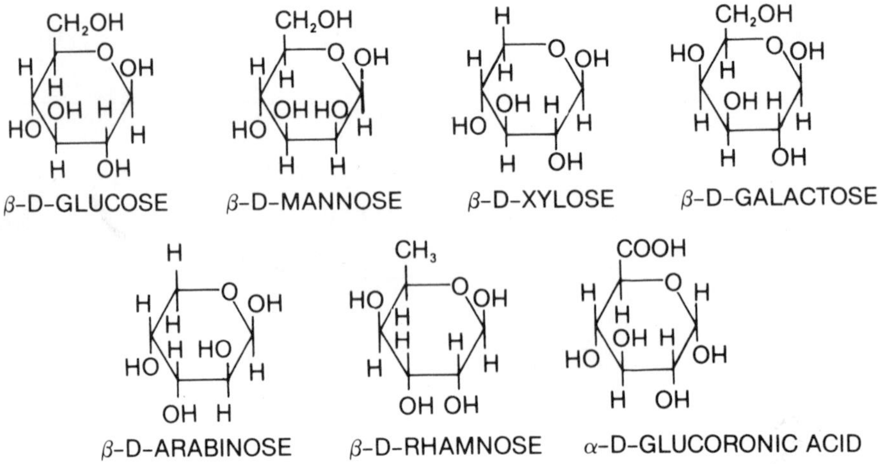
β-D-GLUCOSE
β-D-MANNOSE
β-D-XYLOSE
β-D-GALACTOSE
β-D-ARABINOSE
β-D-RHAMNOSE
α-D-GLUCORONIC ACID

FIG. 2-6. Monomers of hemicellulose. With respect to these monomers, note the presence of such functional groups as methyl and carboxyl (from Wenzl, 1970).

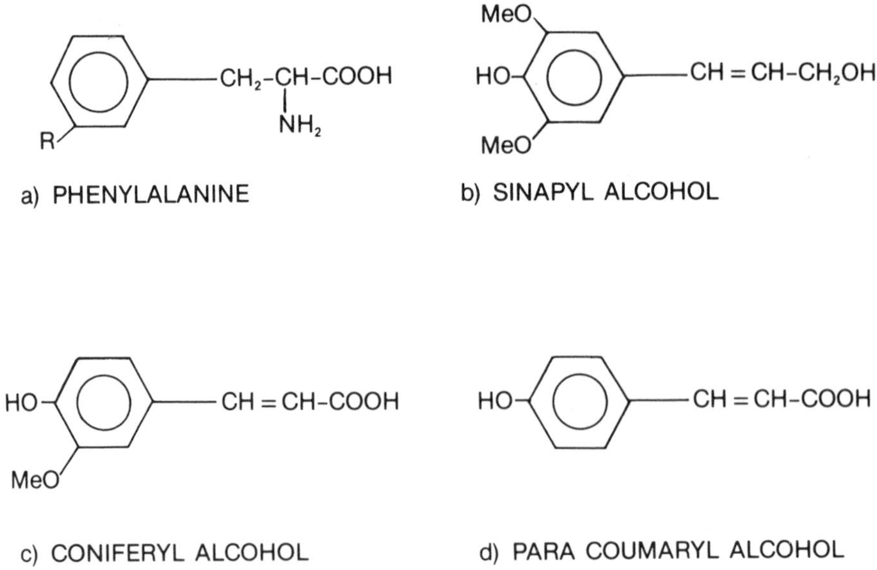
CH₂-CH-COOH
NH₂
R
a) PHENYLALANINE
MeO
HO
MeO
CH = CH-CH₂OH
b) SINAPYL ALCOHOL
HO
MeO
CH = CH-COOH
c) CONIFERYL ALCOHOL
HO
CH = CH-COOH
d) PARA COUMARYL ALCOHOL

FIG. 2-7. Monomers of lignin. With respect to these monomers, note the introduction of phenal rings to the wood structure, the presence of an amine functional group in phenylalanine, and the presence of methoxy functional groups (from Sarkanen and Ludwig, 1971).

are very susceptible to chemical attack, particularly pyrolysis
and oxidation.

The major monomeric units of the hemicelluloses are shown in
Fig. 2-6. The hemicelluloses are usually associated with the
cell wall and are readily degraded. While cellulose is a linear
polymer, the hemicelluloses can form branch structures. The
xylans form a β-1-4 linkage and a branched β-1-2 linkage, while
the galactan group form a β-1-3 linkage and a branched β-1-6
linkage, with the uncommon form a β-1-4 linkage. Functional
groups associated with hemicelluloses include methyl, carboxyl,
and hydroxyl units.

B. Lignin

Lignin, often considered the "glue" holding the wood structure together, is composed mainly of phenylpropane units linked
together by various means (Overend, 1979). The precursor of
both softwood and hardwood lignin is phenylalanine, shown in
Fig. 2-7a. It also forms other proteins in the wood. This ac-
counts for some of the nitrogen content in wood. This nitrogen
is readily accessible and is not bound into ring structures.
The phenylalanine, through hydroxylation and methylation, forms
suragyl alcohol (Fig. 2-7b), the major structural unit of hardwood lignin, and coniferyl alcohol (Fig. 2-7c)--the major structural unit of softwood lignin. Para-coumaryl alcohol (Fig.
2-7d) is also formed and occurs in small amounts in softwood and
hardwood lignin (Sarkanen and Ludwig, 1971).

Typically, softwood lignin contains 95 percent coniferyl and
5 percent paracoumaryl alcohol, while hardwoods contain 25-75
percent suragyl alcohol, with the balance being coniferyl and
paracoumaryl alcohol. These lignin precursors undergo resonance-stabilized radical formation after the enzymatic abstrac-
tion of a hydrogen radical from the phenyl OH group. Free radical formation is stabilized by resonance, forming five different

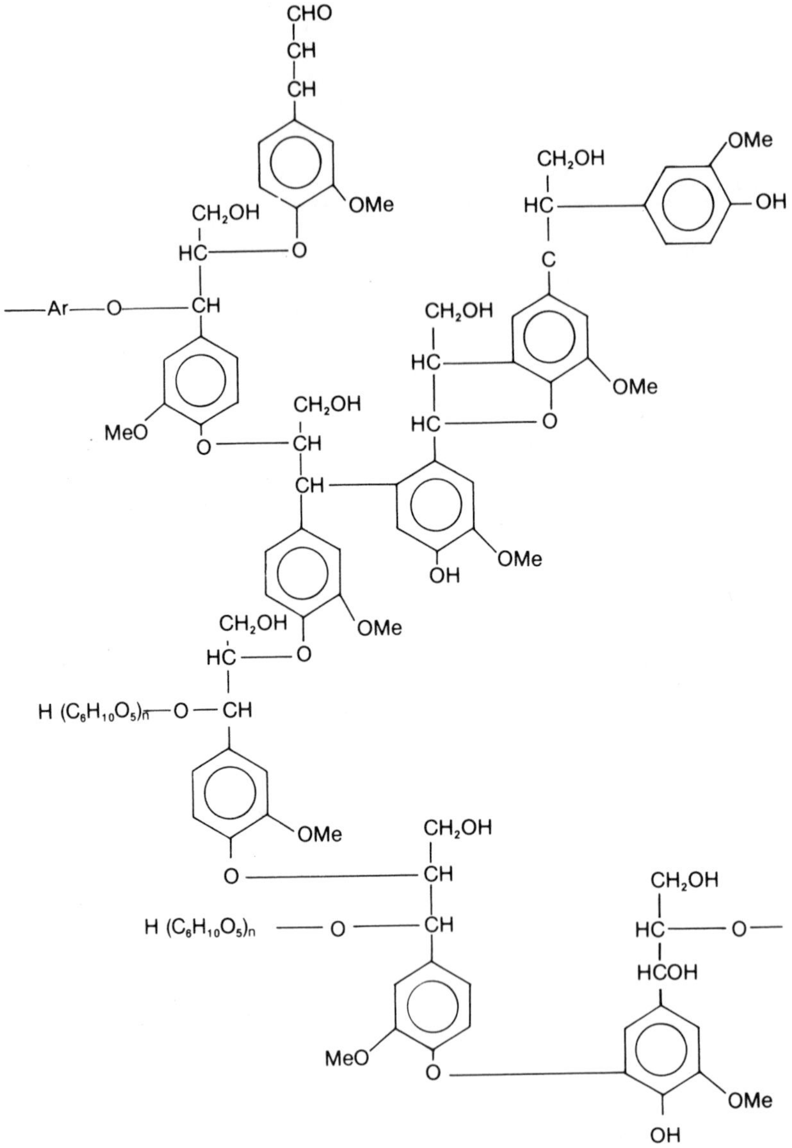

FIG. 2-8. Representative partial structure of softwood lignins. Note that benzene rings are not clustered. Note also the large numbers of methoxy functional groups and the dominance of the β-0-4 linkage between individual units (from Sarkanen and Ludwig, 1971).

free radical structures, which can bond in many ways due to electron rearrangement. The β-0-4, β-5, and β-β are the most common, with the 5-5, β-1, and 4-0-5 being less common (Sarkanen and Ludwig, 1971).

Figure 2-8 is a theoretical structure of lignin, showing bonding to other lignin and cellulose macromolecules. The lignin is of a very complex nature, but it is susceptible to attack.

In comparison to wood, Fig. 2-9 shows a representative partial speculative structure of bituminous coal, indicating the clusters of benzine rings (Sliepcevich et al., 1977). As rank decreases from anthracite to lignite, the degree of clustering also decreases from >3 to 1. Thus, anthracite is very unreactive, while the reactivity of lignite approaches the reactivity of wood. In Fig. 2-9 it is useful to note that the nitrogen appears in both ring and amine form.

C. Proximate and Ultimate Analysis

The proximate anlaysis is a measure of the volatile matter, fixed carbon, and ash content of a given fuel. Table IV presents the proximate analysis of wood, bark, and coal. Coal has substantially less volatile matter while having more fixed carbon ash than wood and bark. The condensed structure of coal imparts increasing stability with rank as can be seen from Figs. 2-6 to 2-9 and Table VI. The large amount of volatile matter associated with wood is a result of the high number of functional groups and the low number of aromatic structures in the wood, as the figures show. This combustion makes wood most reactive when compared to coal. Wood has substantially less ash content than coal. The typical ash analysis of wood is shown in Table V, with major constituents consisting of silicon dioxide (SiO_2) and calcium oxide (CaO).

Table VI presents the ultimate analysis and high heating values for selected fuels. The composition of wood fuels is

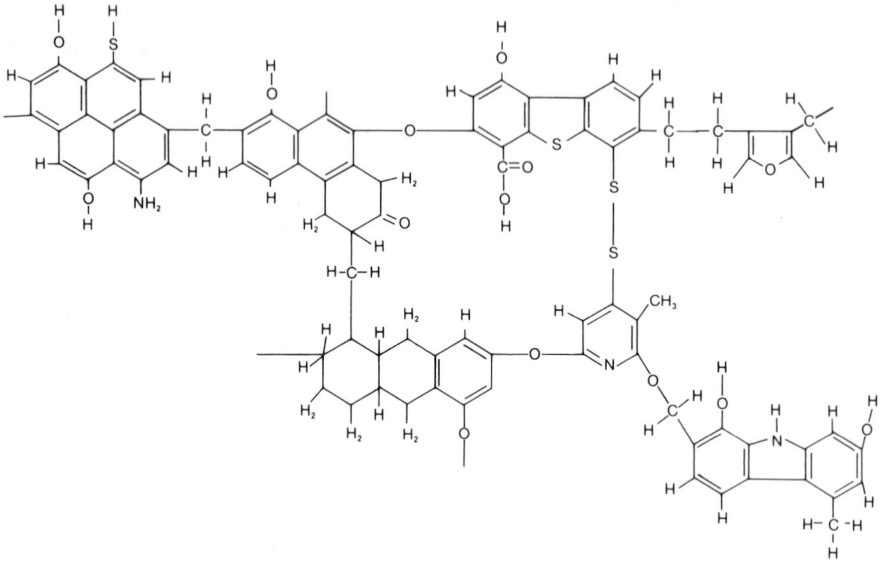

FIG. 2-9. A partial speculative structure of bituminous coal. This structure provides a strong contrast between wood and coal. Note the concentration of benzene rings into clusters of two and three. Note also that nitrogen exists both as amine functional groups and within ring structures. Oxygen exists in some hydroxyl functional groups and a few ether linkages (from Sliepcevich et al., 1977).

fairly constant, with almost no sulfur and a modest amount of nitrogen. Coal also has substantially larger amounts of sulfur and nitrogen. The low oxygen content of coal reflects its increased higher heating value. Peat is shown in order to indicate its intermediate nature between lignite and wood.

TABLE IV. PROXIMATE ANALYSIS OF SELECTED FUELS[a,b]

Fuel	Volatile matter	Fixed carbon	Ash
Wood			
cedar	77.0	21.0	2.0
Douglas fir	86.2	13.7	0.1
hemlock	84.8	15.0	0.2
white fir	84.4	15.1	0.5
ponderosa pine	87.0	12.8	0.2
Bark			
Douglas fir	73.0	25.8	1.2
hemlock	74.3	24.0	1.7
cedar	86.7	13.1	0.2
alder	74.3	23.3	2.4
Coal			
anthracite	6.4	81.4	12.2
bituminous low			
volatile	17.7	71.9	10.4
subbituminous B	40.7	54.4	4.9

[a]Data for wood and bark from Mingle and Boubel (1968), coal from Babcock and Wilcox (1978).
[b]Weight percent, dry basis.

The ultimate analysis can be used to calculate the empirical formula of any fuel, as Table VII illustrates. This formula is then employed to develop combustion equations, to calculate the quantities of air required for combustion, and predict airborne emissions. Table VIII provides an empirical formula for a variety of wood and coal fuels.

The empirical formula can be related to fuel reactivity through the molar hydrogen/carbon (H/C) ratio. This reactivity

TABLE V. ASH ANALYSIS OF WOOD BARK AND COAL[a]

Ash analysis (percent) by weight	Bark				Coal	
	Pine	Oak	Spruce	Redwood	Western	PA
SiO_2	39.0	11.1	32.0	14.3	30.7	49.7
Fe_2O_3	3.0	3.3	6.4	3.5	18.9	11.4
TiO_3	0.2	0.1	0.8	0.3	1.1	1.2
Al_2O_3	14.0	0.1	11.0	4.0	19.6	26.8
Mn_3O_4	Trace	Trace	1.5	0.1	--	
CaO	25.5	64.5	25.3	6.0	11.3	4.2
MgO	6.5	1.2	4.1	6.6	3.7	0.8
NA_2O	1.3	8.9	8.0	18.0	2.4	2.9
K_2O	6.0	0.2	2.4	10.6		
SO_3	0.3	2.0	2.1	7.4	12.4	2.5
Cl	Trace	Trace	Trace	18.4		

[a]From Babcock and Wilcox (1978).

relationship largely results from the presence or absence of aromatic structures discussed earlier. In this regard it is useful to compare H/C ratios of coal and wood. Coal has a molar hydrogen/carbon ratio <1. The H/C ratio increases from anthracite to lignite, finally reaching 1.4 in wood. As will be seen in Chapter 5, this reactivity influences the rate of heat release.

Ultimate analysis, without higher heating value, can be used to predict the latter value by the following formula (Tillman, 1980):

$$HHV_D = 0.475C - 2.38$$

(2-17)

TABLE VI. ULTIMATE ANALYSIS OF SELECTED FUELS[a,b]

Fuel	C	H	O	N	S	Ash	MJ/kg	(Btu/lb)
Softwoods								
Douglas fir	52.3	6.3	40.5	0.1	--	0.80	21.05	(9050)
pitch pine	59.0	7.2	32.7	--	--	1.13	24.22	(10,414)
western hemlock	50.4	5.8	41.4	0.1	0.1	2.20	20.05	(8620)
white cedar	48.8	6.4	44.5	--	--	0.37	17.98	(7728)
redwood	53.5	5.9	40.3	0.1	--	0.20	21.03	(9040)
Hardwoods								
poplar	51.6	6.3	41.5	--	--	0.65	19.09	(8206)
white ash	49.7	6.9	43.0	--	--	0.30	19.09	(8206)
white oak	50.4	6.6	42.7	--	--	0.24	18.85	(8105)
Bark								
oak	39.3	5.4	49.7	0.2	0.1	5.30	19.47	(8370)
pine	53.4	5.6	37.9	0.1	0.1	2.90	21.00	(9030)
redwood	51.9	5.1	42.4	0.1	0.1	0.40	19.42	(8350)
Coal								
anthracite	82.1	2.3	2.0	0.8	0.6	12.20	30.84	(13,258)
bituminous[a]	80.1	4.3	2.2	1.3	1.7	10.40	32.53	(13,986)
subbituminous	72.2	4.8	16.2	1.5	0.4	4.90	28.48	(12,458)
lignite A	68.0	4.6	19.5					
Peat, high decomposed	57.0	5.8	36.0	0.95	0.2	--	--	

[a]Data for coal from Babcock and Wilcox (1975), for peat from Fuchsman (1980), for wood from Pingrey (1976).

[b]Weight percent, dry basis.

[c]Bituminous is low volatile, percent by weight.

TABLE VII. USE OF ULTIMATE ANALYSIS IN CALCULATING THE EMPIRICAL FORMULA OF DOUGLAS FIR BARK[a,b]

Element	Weight (percent)	Molecular weight	gram-moles
Carbon	54.1	12	4.5
Hydrogen	6.1	1	6.1
Oxygen	38.8	16	2.4
Nitrogen	0.17	14	0.01
Sulfur	0	32	0.0
Ash 1.0	N/A	N/A	

[a] Ultimate analysis from Tuttle (1978).

[b] Empirical formula, $C_{4.5}H_{6.1}O_{2.4}N_{0.01}$; basis, 100 g dry wood.

where HHV_D is higher heating value in MJ/kg dry matter and C is percentage carbon. The equivalent of this expression in the English system is (Tillman, 1978)

$$HHV_D = 1.88.0C - 131.5 \qquad (2-18)$$

where HHV_D is expressed in Btu/lb. One can then combine physical and chemical properties to develop a net heating value relationship from expression (2-17):

$$NHV = (0.475C - 238)[1 - (MC/100)] \qquad (2-19)$$

where NHV is net heating value, C percentage carbon expressed on a dry basis, and MC moisture, percentage (green) basis.

The ultimate analysis, then, provides an empirical tool to use in approaching combustion problems.

Based on the above chemical data, wood fuels can be characterized as highly reactive and modest in heating value. These properties significantly influence the combustion characteristics of this fuel.

TABLE VIII. EMPIRICAL FORMULA OF SELECTED WOOD SPECIES[a]

Species	Empirical formula[b]
Softwoods	
Douglas fir	$C_{4.4}H_{6.3}O_{2.5}N_{tr}$
pitch pine	$C_{4.9}H_{7.2}O_{2.0}N_{tr}$
western hemlock	$C_{4.2}H_{5.8}O_{2.6}N_{tr}$
white cedar	$C_{4.2}H_{6.4}O_{2.8}N_{tr}$
Hardwoods	
poplar	$C_{4.3}H_{6.3}O_{2.6}N_{tr}$
white ash	$C_{4.1}H_{7.0}O_{2.7}N_{tr}$
Bark	
oak	$C_{3.3}H_{5.4}O_{3.1}N_{tr}$
pine	$C_{4.5}H_{5.6}O_{2.4}N_{tr}$
redwood	$C_{4.3}H_{5.1}O_{2.7}N_{tr}$
Coal	
anthracite	$C_{6.8}H_{2.3}O_{.12}N_{.06}$
bituminous (low volatile)	$C_{6.7}H_{4.3}O_{.14}N_{.09}$
subbituminous	$C_{6.0}H_{4.8}O_{1.0}N_{.11}$
Peat	$C_{4.7}H_{5.8}O_{3.0}N_{tr}$

[a] Moisture-free basis.
[b] tr: trace amount.

V. CONCLUSION

Wood fuels can be described as hygroscopic and anisotropic materials. Typical moisture contents are approximately 50 percent, specific gravities (dry) are approximately 0.45, and higher heating values are approximately 21 MJ/kg (9000 Btu/lb). Wood fuels are highly reactive, oxygenated, and low in such problematical elements as sulfur and nitrogen.

Because wood fuels are derived from living matter, and because they are produced through a variety of mechanical processes, their combustion characteristics are quite distinct. These issues are addressed in Chapters 3 to 5.

REFERENCES

Babcock and Wilcox (1978). "Steam: Its Generation and Use." The Babcock and Wilcox Co., New York.

Cowling, E. B., and Kirk, T. K. (1976). Properties of cellulose and lignocellulosic materials as substrates for enzymatic conversion processes, in "Enzymatic Conversion of Cellulosic Materials: Technology and Applications" (E. G. Gaden, M. H. Maudels, E. T. Reese, and L. A. Spano, eds.), pp. 95-124. Wiley (Interscience), New York.

Dunlap, F. (1912). The specific heat of wood, in "Water in Wood." Syracuse Wood Science Series No. 4, Syracuse Univ. Press, Syracuse, New York.

Fuchsman, C. H. (1980). "Peat: Industrial Chemistry and Technology." Academic Press, New York.

Kellogg, R. W., and Wangaard, F. F. (1969). Variation in the cell-wall density of wood. Wood and Fiber 1:180-204.

Mingle and Boubel (1978). Data cited in Pingrey (1976).

Overend, R. (1979). Wood gasification: An Overview, in "Hardware for Energy Generation in the Forest Products Industry," pp. 107-119. Forest Products Research Society, Madison, Wisconsin.

Panshin, A. J., and de Zeeuw, C. (1970). "Textbook of Wood Technology," Vol. 1. McGraw-Hill, New York.

Pingrey, D. W. (1976). Forest products overview, in "Energy and the Wood Products Industry," pp. 1-14. Forest Products Research Society, Madison, Wisconsin.

Sarkanen, K. V., and Ludwig, C. H. (1971). "Lignins: Occurrence, Formation, Structure, and Reactions." Wiley (Interscience), New York.

Shafizadeh, F., and DeGroot, W. F. (1976). Combustion characteristics of cellulosic fuels, in "Thermal Uses and Properties of Carbohydrates and Lignins (F. Shafizadeh, K. V. Sarkanen, and D. A. Tillman, eds.). Academic Press, New York.

Siau, J. F. (1971). "Flow in Wood." Syracuse Wood Science Series No. 1, Syracuse Univ. Press, Syracuse, New York.

Skaar, C. (1972). "Water in Wood." Syracuse Wood Science Series No. 4, Syracuse Univ. Press, Syracuse, New York.

Sliepcevich, A., et al. (1977). "Assessment of Technology for the Liquefaction of Coal." The National Research Council, National Academy of Sciences, Washington, D.C.

Stamm, A. J., and Loughborough, W. K. (1935). Thermodynamics of the swelling of wood. J. PHys. Chem. 39:121-132.

Tillman, D. A. (1978). "Wood as an Energy Resource." Academic Press, New York.

Tillman, D. A. (1980). Fuels from waste, in "Kirk-Othmer Encyclopedia of Chemical Technology," Vol. 2, 3rd ed. Wiley, New York.

Tuttle, K. L. (1978). Test report on OSU/Weyerhaeuser experiment. Unpublished.

U.S. Forest Products Laboratory (1974). "Wood Handbook: Wood as an Engineering Material." USGPO, Washington, D.C.

Wenzl, H. (1970). "The Chemical Technology of Wood." Academic Press, New York.

CHAPTER 3

WOOD FUEL SUPPLY: PROCESS CONSIDERATIONS

I.   INTRODUCTION

While spent pulping liquor currently supplies more energy than any other wood fuel, it is almost fully utilized and will supply a decreasing percentage of wood energy in the future. Consequently, the analysis in this chapter focuses on the available marginal increments of wood fuel--mill and forest residues.

Mill and forest residues are available in varying quantities in different regions of the U.S., reflecting their different origins and methods of production. In general, forest residues are widely available, while mill residues, which are available in some areas, are often fully committed to energy. Along with differences in the quantity of fuel available, the quality also varies. This variability in quantity and quality can be traced to the properties of the fuel and to the means by which the fuel is generated.

The most important properties of wood fuel are size, heat content, moisture content, ash content, and the relative surface

48

to volume ratio. These are determined by the system used to produce, collect, and transport the fuel. The most important fuel supply systems can be categorized by the type of fuel and include mill residues, forest residues fuels, and silvicultural fuel farm materials.

II. MILL RESIDUE GENERATION

Three types of mills produce fuels: sawmills, plywood mills, and pulp mills. Of these, sawmills are the most important as they generate more product than plywood mills (Resch, 1978; Boyd et al., 1976; Bethel et al., 1976). Moreover, larger quantities of residues are produced in sawmills than in plywood or pulp mills. Because of their importance, the sawmills are treated extensively here and other types of mills are discussed only by way of comparison.

A. The Sawmill as Fuel Producer

Typical sawmill outputs include (on a weight percent basis) lumber (39), coarse residue for pulp mills (26), sawdust (13), planer shavings (9), and bark (13) (Resch, 1978). Of these, sawdust, planer shavings, and bark are used for fuel. Lumber and coarse residues have higher commodity values. The characteristics of these fuels are shown in Table I.

Table I illustrates the variability in fuel quality alluded to previously. This variability is largely a function of the mill's operating characteristics. For example, logs of a similar size, age, and species will yield significantly different fuel mixtures when processed in different mills. A detailed examination of these factors provides useful information on residue generation.

TABLE I. CHARACTERISTICS OF SAWMILL BASED HOG FUEL[a]

Component	Size range (cm.)	Moisture content (wt percent)	Ash content (wt percent)	Relative surface to volume ratio
Bark	.08-10.12	5-75	1.0-20.0	1
Sawdust	.08-1.0	25-40	0.5-2.0	6
Planer shavings	0.8-1.30	16-40	0.1-1.0	5

[a]From Junge (1975) and Resch (1978).

TABLE II. EQUIPMENT BY MACHINE CENTER

Machine center	Equipment options
Log storage	Dry deck
	Freshwater pond
	Saltwater pond
Debarking	Ring debarkers
	Rosserhead debarkers
	Hydraulic debarkers
Headrigs	Bandsaws
	Circular saws
	Chipping-type headrigs
	Scragg saws
Finishing operations	Dry kiln (optional)
	Planers

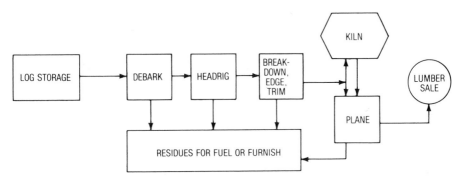

FIG. 3-1. Conceptual representation of a sawmill with emphasis on residue production. With respect to the residues generated by this sawmill, chips would likely go to pulp mills,
planer shavings may or may not be diverted to particleboard
plants or dry-fuel-burning systems, and sawdust may or may not
be diverted to be used in cattle feedlots.

Figure 3.1 is a simplified schematic of a sawmill, showing
the placement of log storage, debarking, headrig and saws, and
finishing activity centers. Table II is a listing of some major
types of sawmill equipment for these activities. Information in
this table is generalized, as indicated by the fact that storage
is identified as a machine operation. This generalization is
compounded by the fact that all equipment beyond the headrig is
lumped into the "finishing" category, with kiln drying shown as
the only real option. Although this approach is oversimplified,
for the purpose of analyzing fuel generation, it provides an
adequate analytical tool. The key operations discussed below
are log storage, debarking, headrig, and drying.

1. Log Storage

Logs may be stored in fresh- or saltwater ponds, or on paved
or unpaved dry decks. Some decks have sprinkler capacity for
maintaining the moisture content of the log, while others do

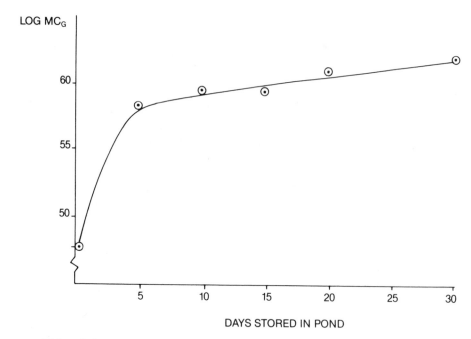

FIG. 3-2. Moisture pickup in wood as a function of log storage in fresh or salt water. Note that moisture pickup is extremely rapid in the first 5 days and then becomes much less significant as the moisture content approaches 60 percent.

not. As the first activity in the milling process, initial storage and handling affects moisture and ash values at all subsequent centers.

Pond storage is characterized by rapid pick-up in the first five days and slower moisture pick-up in subsequent days. Figure 3-2 shows that one dry tonne (t_{od}) of logs typically carries 850 kg of water when it enters log storage and within five days the moisture content is approximately 1.4 tonne of water. At the end of 30 days, water, content may be approximately 1.8 tonnes. The bark on the log increases in moisture content to levels significantly higher than those associated with the wood; and typically 1 tonne of bark will be saturated with 233 t water

(Hardy, 1974) to 300 tonnes of water (Leman, 1976). Essentially the maximum moisture content (MC_{max}) is approached.

Water storage also influences ash content by washing out some of the grit picked up during harvesting (Williston, 1976). Saltwater storage, however, results in some ash pick-up in the form of salt. Hardy has observed salt additions (reported as ash) of 2.5 t per 1500 tonnes of 60 percent MC_g (600 tons of fuel) hog fuel burned. This represents an increase of 0.4 percent in the total weight of the fuel. Also significant is the fact that salt adds to the corrosive properties of the ash. These properties together with the added moisture resulting from water storage characterize water-stored fuels.

Water-sprinkled deck storage results in some moisture pickup in the bark, although far less water is absorbed in water storage. For example, whereas moisture contents of 165 percent are common in logs stored under water for 30 days, Miller and Swan (1980) show average moisture content values of less than 62 percent (160 percent MC) in sapwood stored on sprayed decks for 47 days. Further, 15 percent of the moisture content values are approximately 50 percent (100 percent MC) in this study of ponderosa pine storage. It is doubtful that, under this system, unacceptably high maximum moisture contents MC_{max} would be approached.

Dry storage provides no mechanism for removing ash and grit acquired during harvesting, and unpaved yards may add more grit (ash) to the fuel. Values of up to 5 percent ash have been found for bark (Junge, 1975), with much of this attributable to grit and sand picked up on the dry deck.

Water storage of logs is being reduced and dry deck storage is being increased. This trend promises modest improvements in the moisture content of hog fuel being burned, while suggesting that ash quantities will increase. This trend will reduce the number of mechanical press bark dryers that are presently used in order to produce a burnable fuel.

2. Debarking

Debarking systems in the sawmill include hydraulic jets, rink debarkers, and rosserhead debarkers. The hydraulic system uses water pressure to tear chunks of bark from the log, while the latter two systems employ mechanical action. The system employed influences the moisture content, the grit content, the particle size distribution, and the volume of fuel itself.

The influence of mechanical and hydraulic debarking on moisture content has been described by G. W. Christianson (personal communication, November 15, 1977). Mechanical debarking produces fuel with MC_g ranging from 40 to 50 percent and a mean of approximately 45 percent. Hydraulic debarking results in a fuel with MC_g = 67 percent. Mechanical debarking appears to remove some moisture while hydraulic adds water to the fuel. The influence on ash and grit content was also discussed by Christianson. Ash contents of bark are 4-5 percent for fuel from mechanical debarking systems and 2-3 percent for fuel from hydraulic systems. Mechanical systems yield 27 percent bark at less than 1/4 in. diam, 47 percent between 1/4 and 4 in., 26 percent at >4 in., and 17 percent at 1 ± 0.25 in. Hydraulic systems yield 70 percent of the bark at <1/4 in., 1 percent at >4 in., and 19 percent at ~1 in.

Along with these qualitative differences, the influence of debarking also affects the quality of various fuels produced. Ring and hydraulic debarking systems tend to produce residues with 98.7 percent bark and 1.3 percent wood (weight basis) while rosserhead and drum debarkers produce residues of 89.7 percent bark, and 10.3 percent wood (Searles, 1969).

3. Headrigs and Saws

Headrigs, ponys, resaws, edgers, and trimmers play another substantial role in fuel generation. Given the differing amounts of wood available to these saws plus variations in sawdust generation, more heterogeneity is introduced into the

TABLE III. PRODUCTS OF SAWMILLS BY HEADRIG TYPE[a,b]

Headrig type	Lumber	Coarse (pulpable) residue	Sawdust fuel
Chipping	55-56	38-41	4-6
Circular saw	59-61	20-27	14-19
Band saw	60	26-28	12-14
Scragg saw	50-51	24-30	20-25
Log gang saw	56	32	12

[a]From Hartman et al. (1978).

[b]Percent dry weight basis. Assumes debarked logs.

fuel. Table III, based on Hartman et al. (1978), shows the influence of select headrigs on quantities of fuel generated.

Unlike the data showing the quantities of fuel generated, the literature does not provide data on moisture content as a function of headrig type. Junge (1975) identifies a range in MC_g = 25-40 percent (green) for sawdust in hogged fuel. Jamison (1979) gives values of 40-55 percent for the same parameters. These variations are not surprising, given a feedstock MC_g = 33-55 percent (45-60 for softwood) and the variation in water used for cooling sprays. In light of these uncertainties and in the absence of a better procedure, MC_g = 5 percent can be added arbitrarily to the as-received condition of the wood. As a result of the operation, dry-stored softwood logs yield sawdust with MC_g = 50 percent, well within the range posited by Jamison. Hardwoods stored dry have a sawdust MC_g ~ 33 percent, consistent with both the Jamison and Junge data.

4. Planing and Dry Kiln Operations

Lumber may be rough and green; planed and sold; or dried, planed, and sold. The difference is not in the quantity of residue produced but in the moisture content of the fuel stream. Volumes are taken at approximately 10 percent of the sawn, trimmed, and edged lumber. Moisture contents of 40 percent are associated with green shavings (Jamison, 1979), and 15 percent for dry shavings (G. W. Christianson, personal communication, November 15, 1977). On this basis one can calculate the quantity of dry or green fuel produced as a function of the process employed.

5. Sawmill Fuel Production

The best way to summarize the effect of various processes on fuel variability is to construct a series of sawmill process trains, and examine the combined influence of these systems on fuel volume, particle size distribution (percent fines), moisture content, and ash content. Figure 3-3 graphically depicts one sawmill processing scheme. Table IV summarizes fuel values associated with three different processing systems. As indicated in Figure 3-3 and Table IV, it is apparent that hog fuel is wet, loaded with fine particles, and subject to wide (relative) swings in ash content.

The extent of variation is shown in Fig. 3-4. This Figure contains two photographs of hog fuel samples obtained in the Black Hills. Figure 3-4a illustrates the range of particle sizes. Figure 3-4b shows the total composition including the presence of many fines.

It should be observed that the values in Table IV are not complete. For example, the fuel entering the boiler is not shown in the table. Thus, factors such as weather and fuel storage characteristics, which influence the quality of wood fuels as they enter utilization systems, are not readily analyzed. Further, the values posited are approximate averages,

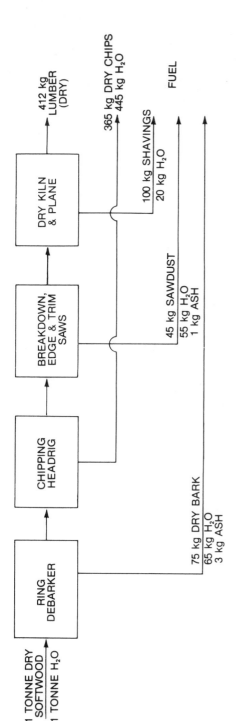

FIG. 3-3. Fuel production from one sawmill flow sheet. With respect to this flow sheet, note that the presence or absence of dry planer shavings may change the moisture content of the resulting fuel, in total, by 10 percent. It may also change the ash content, on a dry basis, by 1.4 percent. As processes are changed to include different types of debarkers, headrigs, and finishing operations, fuel composition changes as well.

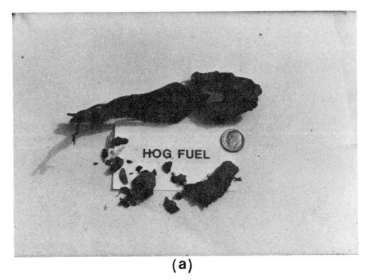

(a)

(b)

FIG. 3-4. Hog fuel from a typical sawmill: (a) individual
particles of this hog fuel, illustrating the particle size variation associated with fuels from sawmills; (b) a composite of
such fuels. A dime is inserted in each photograph for perspective. With respect to these photographs, note the influence of
processing schemes such as that shown in Fig. 3-3 on the quality
of fuel produced. Photographs courtesy of Envirosphere.

TABLE IV. CALCULATED HOG FUEL VALUES AS A FUNCTION OF SAWMILL MACHINE CENTERS

Storage debark headrig finishing	Dry, paved ring chipping kiln dry	Dry, unpaved ring circular saw kiln dry	F.W. Pond Rosserhead circular saw plane, no dry

Total fuel pro-
    duced (kg, per
    ton of logs,
    O.D. basis)
	185	265	274

Ash content (percent)	1.6	1.8	1.4
MC_g (percent)	52	51	58

Net heating value
    ($J \times 10^6$
    per kg fuel)
	2.14	2.20	1.88

Fines (<1/4 in.) (percent)	37	54	69

and individual situations will vary from these expected values. Finally, this figure does not reflect the fact that markets resulting from other uses of residues seriously impact fuel composition by removing almost all coarse wood and some sawdust for pulping, some shavings to particle board manufacture, and some bark and sawdust to miscellaneous uses such as landscaping, cattle bedding, and residential fuels (e.g., presto log) manufacture.

B. Plywood Mill and Secondary Manufacturing Residue Fuels

Plywood plant and secondary manufacturing residues constitute another important mill residues used as fuel. In 1976, there were 7.4×10^6 tonnes (O.D.) of plywood residues and 2.0×10^6 tonnes of associated residues produced (Phelps, 1977; Boyd et al., 1975). This compares with 30.5×10^6 t of lumber produced in 1976, along with 18.3×10^6 t of hog fuel in sawmills (Phelps, 1977; Boyd et al., 1976). Additionally, 1.5 tonnes of secondary manufacturing residues were produced in that year (Bethel et al., 1979). These are important sources of fuel, but less significant than those discussed previously.

These fuels are distinguished from sawmill residues by several physical properties, primarily moisture content. In plywood manufacture bark waste is produced in a manner similar to lumber debarking waste. For every tonne of barky log entering the plywood plant, however, 100 kg of bark (O.D.) and 20 kg of dry sawdust are produced (Boyd et al., 1976). Figure 3-5 presents a flow sheet for plywood production, which includes fuel generation. In the furniture mills and other secondary manufacturing plants, virtually all of the residue is dry. Such fuels are higher in net heating value than sawmill hog fuels and can be burned in a wider range of combustion equipment.

These mill residues constitute the most desirable fuel in locations where there is an imbalance in fuel-producing and -consuming facilities. Such imbalances are common in areas where sawmills are abundant but where pulp and paper mills are distant or otherwise provide an undesirable market for chips.

III. FOREST RESIDUE FUELS

Forest residues are generated as a result of forest management activities. Such residues fall into several categories, but it is best to discuss them according to increments believed to correspond with the broad units of a marginal cost curve. The order is tyically logging residues, dead and dying timber, improvement cuttings, and stand conversions. This order corre-

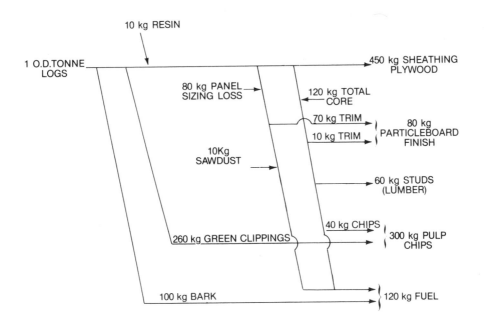

FIG. 3-5. Material balance of a sheathing plywood mill. Note the distribution of residues to particleboard manufacturing, to sawmills, to pulpmills, and to the powerhouse. For every tonne of logs comsumed (dry basis), only 120 kg of fuel are made available to the powerhouse.

sponds not only to a rough marginal cost curve (see Bethel et al., 1979), but also to levels of silvicultural intensity [see Kitto (1980) for a discussion of this point].

A. Logging Residues

Logging residues from natural and managed stands include tops, limbs, and foliage--equal to between 15 and 20 percent of the above-ground dry biomass in the tree. Additional logging residue comes from broken and defective bole-wood materials. Currently, some 75×10^6 tonnes/year of these materials are

TABLE V.   DENSITY OF LOGGING RESIDUES FROM CLEARCUTS IN
WASHINGTON[a,b]

	Forest type	
Region/Owner	Old growth	Young growth
Western Washington		
forest industry	27.2	11.9
national forest	87.6	11.9
other	51.5	11.1
Eastern Washington		
forest industry	41.2	3.8
national forest	41.2	3.8
other	41.2	3.7

[a]From Bergvall et al. (1978).
[b]Values in bone dry tonnes per hectare.

produced (Jamison, 1979), containing 1.5 EJ (1.4 x 10^{15} Btu).
These fuels vary among species, although they are generally wet
with moisture content, averaging about 100 percent.

In natural stands, logging residues come in a wide variety
of sizes and shapes. The residues from managed stands have more
predictable sizes and shapes due to the characteristics of the
stand from which they were produced. Further, in natural old-
growth stands, logging residue production is much higher than in
younger managed stands. Table V illustrates this point showing
logging residues/densities in Western Washington.

Logging residues provide a potential fuel supply for industry. In addition, removal of these residues for use as fuel
provides silvicultural and environmental benefits. From a silvicultural point of view, their removal hastens and enhances
site preparation through site scarification (Bethel et al.,

1979). From an environmental perspective, their removal reduces
the amount of prescribed burning on site, with an attendant reduction in atmospheric emissions of particulates, carbon monoxide, and hydrocarbons.

The net result of these considerations is that while the
quantities of logging residue are expected to decrease over time
as natural stands are replaced by managed stands, their use may
become more widespread as a function of energy price, silvicultural savings, and environmental benefits. A key consideration
affecting the use of these logging residues is the fact that the
piece size of this fuel can be controlled at relatively uniform,
desirable levels (e.g., 1.2-3.7 cm screen size). Problems that
result from other systems because of the presence of fines are
minimized by the use of in-woods chippers or in-woods hogs.
Such a system results in more uniform particle size distribution
than in mill residue systems where debarking and other processing activities leading to the production of fines occur.

B. Stand Improvement Materials

Stand improvement materials include undesirable hardwoods
that typically have a moisture content of approximately 45 percent (Goldstein et al., 1978); juvenile wood precommercially
thinned, with MC_g = 51-60 percent (Britt, 1970); and dead timber with a MC $\cong$ 25 percent (Grantham and Howard, 1980). These
materials represent a resource that is available during the period of conversion from natural to managed stands. Beyond that
period of time, thinnings and a modest amount of mortality will
provide the basis of this fuel increment. Bethel et al. (1979)
estimates the total quantity of material available at 2.2 x-
10^9 t (O.D.) or 43 x 10^{18} J. On a 20 year basis, that is
equivalent to 2.2 x 10^{18} J/yr.

In assessing the characteristics of this fuel, it should be
recognized that this resource is lower in moisture content than

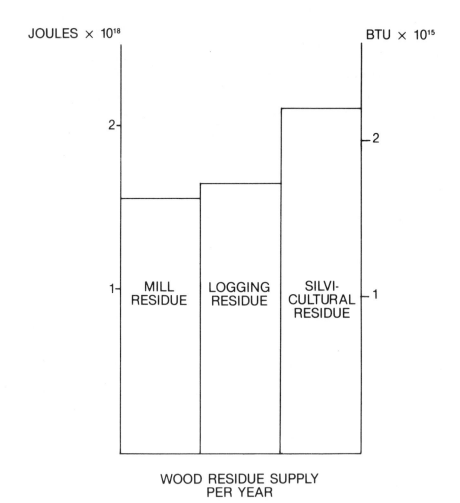

FIG. 3-6. Annual wood residue supply in the United States. This quantity of fuels available will change as new markets for mill residue are developed, as logging operations shift from old growth to second growth, and as silvicultural practices are increased.

mill residues, and that fines generation can be minimized. This fuel, like logging residues, can be generated to a more uniform size and quality than sawmill hog fuel.

Figure 3-6 shows the annual quantities of wood fuel available, using a 20 year consumption rate on unsalable timber (e.g., dead trees, stand conversion materials) and disregarding the price of such fuels. From Fig. 3-6, given projected declines in the production of logging residues and the possible mid-term nature of silvicultural residues, it is important to consider wood residue factors as they affect supply in the long term.

IV. SILVICULTURAL FUEL FARM MATERIALS

The concept of growing trees for use as fuel is not new; it was employed by the iron industry in New Jersey and Pennsylvania at the beginning of the nineteenth century (Walker, 1966). Similarly, the concept of coppice regeneration was practiced in Europe at least a century ago. During the 1970s the concept of growing fuels on energy farms was revived, as solutions to problems of fossil hydrocarbon supplies were sought.

The studies by Intertechnology Corp. (e.g., Szego and Kemp, 1973), as well as the Mitre Report (Inman et al., 1977; particularly Howlett and Gamache, 1977) have laid out the basic concept: that a tract or series of tracts of land would be dedicated to the growth of biomass for use in energy conversion. Biomass species would be selected on the basis of their ability to grow rapidly as soon as planted or in the early years of their lives. Management systems would be intensive, approaching agricultural techniques. Harvesting would occur annually or on a slightly less (2-10 year) frequent basis. The biomass would then be transported to, and used in, industrial facilities.

Numerous forms of biomass have been proposed for these fuel farms including sugar cane, grasses, and trees. Much emphasis has been placed upon certain hardwoods in the development of this concept. Advantages from using hardwoods include the flexibility of year-round harvest not associated with sugar cane or grasses, rapid growth, and coppice regeneration. The last feature reduces fuel farm operation costs. The Mitre Report, in particular, highlighted certain preferred deciduous species.

Critical aspects of silvicultural fuel farms include establishment of the desired tree species on a dedicated land base, managing the growing stock to produce the highest yields possible in a reasonably short period of time, harvesting the fuel crops in an economically efficient manner, and transporting the fuel to the various conversion facilities. The cycle of growth, management, harvest, and transportation must be maintained over an economic life of 25-30 years.

Conceptually, fuel farms represent the adaptation of biological systems to mass production needs. The intent of such systems is use of the forest to serve the fuel needs of conversion systems.

Because residues are not an unconstrained supply system, silvicultural fuel farms have been proposed as a production system capable of adding biomass energy supplies to the system. Although no such system is being used today, its potential merits scrutiny.

The basic concept of the silvicultural fuel farm is to grow trees over short rotations with intensive silviculture. The trees so grown would be harvested specifically for fuel. Key production variables are yields in dry metric tons/hectare/year, silvilcutural regimes, and expenditures (e.g., fertilization) associated with producing and delivery of the fuel. Critical advantages, from a production and utilization perspective, are producing a fuel of relatively uniform piece size, energy content, and moisture content.

A. Silvicultural Yields and Regimes

The energy production of silvicultural fuel farms is dependent on yield estimates for deciduous species. These estimates
are infuenced by the assumptions that the best land (e.g., site
classes I and II) will not be available, as these areas will be
devoted to growing high-value softwoods. Only site class III
and lower-quality land can be expected to be available for fuel
farming. For the Southeast, low to average land (e.g., SI = 60,
25 year basis) is used. For the Pacific Northwest, a site index
of 90 (50 year basis) may be used.

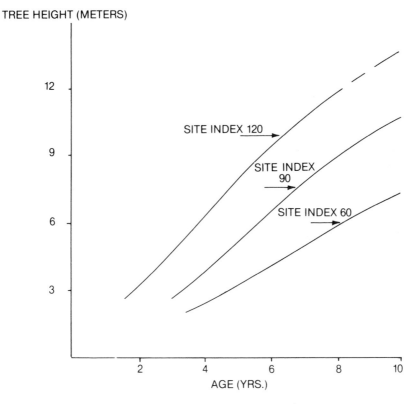

FIG. 3-7. The influence of site quality on tree height for
juvenile red alder (site index basis: 50 years)(from Smith and
DeBell, 1974).

Growth and yield data for hardwoods are generally not well developed over a broad range of species. Thus, on the basis of data available, only American sycamore and red alder are discussed below.

For American sycamore, data are available on site indexes of 82 and 90 (25 year basis) lands (Steinbeck and May, 1972). Belanger and Saucier (1975) have interpreted data of Saucier et al. (1972), and Kormarik et al. (1973), to show that lower-quality sites will yield only 58 percent of the output grown on high-quality lands. That proportion is used here in the absence of more definitive data. On a spacing of 1.2 m x 1.2 m (4 ft x 4 ft), a mean annual increment (MAI) of 1.5 O.D. t/hectare/year (4.2 O.D. tons/acre/year) at five years is assumed for low sites.

The best data on short rotation red alder, which are site sensitive, are presented by Smith and DeBell (1974). These data correlate tree height and age as a function of site quality, as is shown in Fig. 3-7. Yields and mean annual increments are then presented as a function of tree height and spacing. An MAI of 2.6 O.D. t/hectare/year (7.0 O.D. tons/acre/ year) at age 10 is reasonable for site class III and IV land (Tillman, 1980).

In order to obtain such high yields, a regime consisting of extensive site preparation, planting, fertilization, and weed control is necessary. Fertilization at an application frequency of every 5 years on sycamore and every 10 years on alder is required (Tillman, 1980).

B. Harvest and Transport Systems

Fuel farming is contemplated for relatively large (e.g., 4000 hectare) tracts. Thus, all operations are expected to be highly mechanized. Typical harvesting systems contemplated are either the presently available feller-buncher and wheel-skidder, or the mobile Koch-type chippers now in research and development. Such harvesting systems typically can cut 5 stems/minute (J. B. Shelby, Weyerhaeuser Company, personal communication, November 15, 1979), or cover 0.4 hectares/hr, removing 23 O.D. tonnes of biomass.

Chippers, either of the Morebark 380 hp design or the Koch design, typically can process 20-23 O.D. tonnes/hr. Thus, crews can be balanced for the production of fuel farm biomass systems. Chippers typically load chip vans or receiving hoppers directly.

Transportation systems associated with fuel farms normally are 25 tonnes payload over-the-road chip vans. These systems are common in the pulp and paper industry and generate a minimum quantity of fines in the hauling of wood chips.

C. Fuel Quality

As has been mentioned previously, fuel farm grown materials raised in monocultures have the potential for being more uniform than mill residues. Specific gravity may be lower than expected due to the juvenile nature of the wood; for American sycamore a value of 0.46 may be expected (Zobel, 1980). MC_g may be higher, exceeding 55 percent (Zobel, 1980). Ash content is governed partially by harvesting, as previously discussed in mill residues. Piece size is governed by chipping equipment; and large pieces (e.g., 1-2.4 cm) can be expected as operators minimize the energy requirements of their harvesting equipment.

These fuel composition values may or may not be desirable when compared to some residue fuel characteristics. Moisture content of some mill residues, for example, is approximately 20 percent compared to the 55 percent for juvenile hardwood. However, uniformity is an aid to the utilization system design. Further, the degree of uniformity can only be confirmed by the operation of one or more such fuel farms. Whether such installations are ever built is largely a function of the cost of such fuels, particularly as such costs related to residuals and fossil fuels.

V. CONCLUSIONS

From the data presented above it is clear that mill residues, while most accessible, exhibit the greatest variation in particle size, moisture content, ash content, and other relevant physical and chemical parameters. This variation is caused by specific processes used to manufacture dimension lumber and other solid wood products. As one moves from mill residues to forest residues, and from all residues to materials grown specifically for fuel purposes, variability in critical fuel parameters can be brought under increasing control.

The data above also demonstrate that quantities of potentially availabe wood fuel are not unlimited. Mill residues and logging residues are produced in response to demand for wood products: lumber, plywood, and pulp. Similarly, silvicultural residues are produced as a function of investments in stand conversion and thinning largely by the forest products industry and, to a lesser extent, by other private land holders and the U.S. Forest Service. With silvicultural fuel farms the physical supply of wood fuels can be increased.

The data above are limited to issues of physical and chemical characteristics, and physical availability. Implications of those characteristics on the utility of wood fuels, and economic availability of those wood fuels, are issues to be addresssed in subsequent chapters.

REFERENCES

Belanger, R. P., and Saucier, J. R. (1975). Intensive culture of hardwoods in the South. Iowa State J. Res. 49:339-344.

Bergvall, J., et al. (1978). "Wood Waste for Energy Study--Inventory Assessment and Economic Analysis 1978." State of Washington, Dept. of Natural Resources, Olympia, Washington.

Bethel, J. S., et al. (1976). "The Potential of Lignocellulosic Material for the Production of Chemicals, Fuels, and Energy." National Academy of Science (Committee on Renewable Resources for Industrial Materials), Washington, D.C.

Bethel, J. S., et al. (1979). "Energy from Wood." A Report to the Office of Technology Assessment, Congress of the United States. College of Forest Resources, Univ. of Washington, Seattle, Washington.

Boyd, J., et al. (1975). "Problems and Legislative Opportunities in the Basic Materials Industries." National Research Council, National Academy of Sciences, Washington, D.C.

Boyd, C. W., Koch, P., McKean, H. B., Morschauser, C. R., Preston, S. B., and Wangaard, F. F. (1976). Wood for structural and architectural purposes. Wood and Fiber 8.

Britt, K. W. (1970). "Handbook of Pulp and Paper Technology." Van Nostrand-Reinhold, New York.

Goldstein, I. S., Holley, D. L., and Deal, E. L. (1978). Economic aspects of low grade hardwood utilization. Forest Prod. J. 28.

Grantham, J. B., and Howard, J. O. (1980). Logging residue as an energy resource, in "Progress in Biomass Conversion," Vol. 1 (K. V. Sarkanen and D. A. Tillman, eds.), pp. 1-35. Academic Press, New York.

Hardy, H. J. (1974). Drying hogged fuel for powerhouse operations, in "Modern Sawmill Techniques," Vol. 3. Freeman, San Francisco.

Hartman, D. A., Atkinson, W. A., Bryant, B. S., and Woodfin, R. O. (1978). "Conversion Factors for the Pacific Northwest Forest Industry." College of Forest Resources, Univ. of Washington, Seattle.

Howkett, K., and Gamach, A. (1977). "Silvicultural Biomass Farms, Vol. 2: The Biomass Potential of Short Rotation Farms." The MITRE Corp., McLean, Virginia.

Inman, R. E., et al. (1977). "Silvicultural Biomass Farms, Vol. 4: Site-Specific Production Studies and Cost Analyses." MITRE Tech. Rep. No. 7347, U.S. Dept. of Commerce, Washington, D.C.

Jamison, R. L. (1979). Wood fuel use in the forest products industry, in "Progress in Biomass Conversion," Vol. 1 (K. V. Sarkanen and D. A. Tillman, eds.), pp. 27-52. Academic Press, New York.

Junge, D. C. (1975). "Boilers Fired with Wood and Bark Residues." Res. Bull. 17, Forest Research Laboratory, Oregon State Univ., Corvallis.

Kitto, W. D. (1980). Environmental considerations in wood fuel utilization, in "Progress in Biomass Conversion," Vol. 2 (K. V. Sarkanen and D. A. Tillman, eds.). Academic Press, New York.

Komarik, P. E., Tyre, G. L., and Belanger, R. D. (1973). A case history of two-short-rotation coppice plantations of sycamore of Southern Piedmont bottom lands, in "IUFRO Biomasss Studies" (H. E. Young, ed.). College of Life Science and Agriculture, Univ. of Maine, Orono.

Leman, M. (1976). Air pollution abatement applied to a boiler plant firing salt-water soaked hogged fuel, Proc. FPRS Energy Workshop No. P-75-13.

Miller, D. J., and Swan, S. (1980). Blue stain in sprinkled log decks and lumber piles of Ponderosa Pine. For. Prod. J. 30:39-41.

Phelps, R. B. (1977). "The Demand and Price Situation for Forest Products 1976-77." USDA Forest Service, Misc. Publ. No 1357, USGPO, Washington, D. C.

Resch, H. (1978). "Energy Recovery from Wood Residues." Paper presented at the Eighth World Forestry Congress, Jakarta.

Saucier, J. R., Clark, A., and McAlpine, R. G. (1972). Above ground biomass yields of short rotation sycamore. Wood Sci. 5:1-6

Searles, R. L. (1969). Economic aspects of bark utilization, in "Making and Selling Bark Products," Proc. Forest Prod. Res. Soc. Mtg., Madison, Wisconsin.

Smith, J. H. G., and DeBell, D. S. (1974). Some effects of stand density on biomass of Red Alder. Can. J. Forest Res. 4:335-346.

Steinback, K., and May, J. T. (1972). "Productivity of Very Young Plantanus Occidentalis." IUFRO Biomass Studies, Univ. of Maine, Orono.

Szego, G. C., and Kemp, C. C. (1973). Energy forests and fuel plantations. Chemtech, May 3, pp. 275-284.

Tillman, D. A. (1980). Fuels from waste, in "Kirk-Othmer Encyclopedia of Chemical Technology," Vol. 2, 3rd ed. Wiley, New York.

Walker, J. E. (1966). "Hopewell Village: The Dynamics of a Nineteenth Century Iron-Making Community." Univ. of Pennsylvania Press, Philadelphia.

Williston, E. M. (1976). "Lumber Manufacturing: The Design and Operation of Sawmills and Planer Mills.: Freeman, San Francisco.

Zobel, B. (1980). Genetic improvement of forest trees for biomass production, in "Progress in Biomass Conversion," Vol. 2 (K. V. Sarkanen and D. A. Tillman, eds.), pp. 37-58. Academic Press, New York.

CHAPTER 4

THE PROCESS OF WOOD COMBUSTION

I. INTRODUCTION

The physical and chemical characteristics of wood fuel are largely described in Chapter 2. These characteristics determine the materials handling characteristics of wood fuel. They also determine the pathways and mechanisms associated with wood combustion. Those pathways and mechanisms are the focus of this chapter. Emphasis is placed on combustion of large, wet wood particles. Some comparisons are made between wood and coal combustion.

The conceptual model of solid-fuel combustion proposed by Edwards (1974) facilitates this analysis. In the Edwards model, shown in Fig. 4-1, five reaction zones exist: the nonreacting solid zone, the condensed phase reaction zone, the gas phase reaction (pyrolysis) zone, the primary (gas phase) combustion zone, and the postflame reaction zone. For the purpose of reviewing wood combustion reaction mechanisms, four basic stages can be employed: (1) heating and drying, (2) solid-particle py-

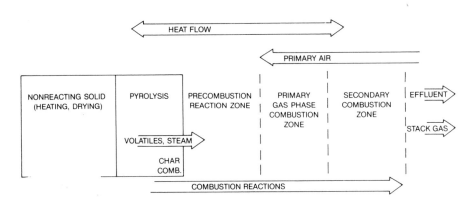

FIG. 4-1. The conceptual model of solid-fuels combustion adapted from Edwards (1974). This model includes numerous reaction zones and helps explain the complexities associated with wood combustion.

---

rolysis, (3) gas phase pyrolysis and oxidation, and (4) char oxidation. Heating and drying occurs in the nonreacting solid zone. Both solid-particle pyrolysis and char oxidation occur in the condensed-phase reaction zone.

This analysis involves an examination of mechanisms associated with wood combustion, without consideration of the kinetics of specific reactions. Strict mathematical analysis could be performed by employing generic equations such as those proposed by Sohn and Szekely (1972). Numerous assumptions would have to be made in such an effort, however, as the fundamental kinetic values are only beginning to emerge. Significant gaps still exist in the fundamental data base (Antal, 1979; Bailie, 1977; Maa and Bailie, 1978). Beyond these problems, the use of a descriptive, mechanistic approach facilitates a consideration of the multifaceted influence of moisture on wood combustion.

Application of this model as a review tool is facilitated by a brief review of the physical and chemical properties of wood discussed in detail in Chapter 2. Physically, wood is both anisotropic and hygroscopic. For softwoods, the main structural component is the tracheid, which varies in length from 3.5 to 7 mm. Tracheids are hollow, but the lumen, the interior void space, can contain water. Moisture migration between tracheids takes place primarily through pit structures, small-diameter valvelike openings. Moisture content may be up to ~23 percent as bound water and up to 75 percent (green basis) as total moisture (Cowling and Kirk, 1976; U.S. Forest Products Laboratory, 1974; Wenzl, 1970). Chemically, wood is composed of polysaccharides (cellulose and hemicelluloses) and lignin, and behaves as a polyalcohol. Cellulose is a linear polysaccharide composed of anhydroglucose units connected by $1 \longrightarrow 4$-β-glucosidic linkage. The principal functional group is the OH group. However, upon oxidation, functional groups will include carbonyls, ketos, and carboxyls. Hemicelluloses are branched-chain polysaccharides. In deciduous woods, the principal component is 4-0-methylglucuronoxylan. In softwoods the principal component is glucomannan.

Functional groups associated with hemicelluloses include carboxyls, methyls, and hydroxyls. Lignin contains a basic skeleton of four or more substituted phenylpropane units. The basic building blocks are quaiacyl alcohol (for softwoods) and syringyl alcohol. Bonding between units is dominated by the β-4 ether linkage with β-5, α-alkylether, 5-5, α-α, β-β, and other linkages existing (Cowling and Kirk, 1976; U.S. Forest Products Laboratory, 1974; Sarkanen, 1971).

Figure 4-2 shows representative partial structures associated with wood fuels. These structures show the highly oxygenated nature of wood fuels and the location of functional groups, and they are highly useful in reviewing pyrolysis and oxidation mechanisms associated with wood fuels. Table I provides a review of key analytical values for wood as fuel.

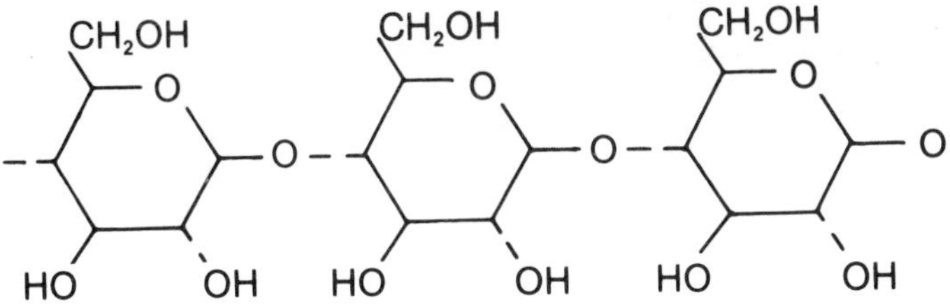

CELLULOSE

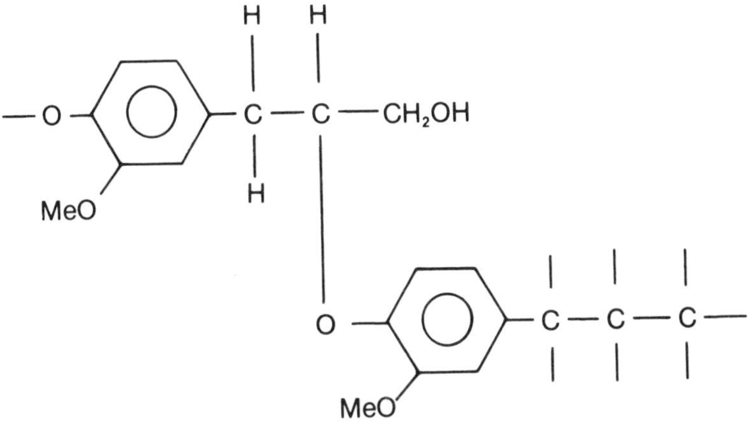

PARTIAL SOFTWOOD LIGNIN STRUCTURE
SHOWING
β-o-4 LINKAGE

FIG. 4-2. Representative structures for holocellulose and lignin in wood fuel. Holocellulose is represented by the cellulose chain. The lignin representation is used to show the β-0-4 linkage.

TABLE I. ESTIMATED AVERAGE COMPOSITION VALUES
FOR SOFTWOODS AND HARDWOODS[a]

	Wood type (percent of weight)	
	Softwoods	Hardwoods
Chemical composition (dry basis)		
Cellulose	43	43
Hemicelluloses	28	35
Lignin	29	22
Ultimate analysis (dry basis)		
H	6.1	6.2
C	53.0	51.0
O	38.8	39.9
S	--	--
N	0.1	0.2
ash	1.7	2.5
Proximate analysis		
volatiles	40.6	52.4
fixed carbon	12.4	12.9
ash	1.0	2.7
moisture	46.0	32.0

[a]From Shafizadeh and Lai (1975), Wenzl (1970), Arola (1976), and Junge (1975).

It is useful to continue the comparison between wood and coal here. Representative model coal structures are shown in Chapter 2. It is significant to note that coal is more aromatic and has few hydroxyl functional groups and ether linkages. Heteroatoms such as nitrogen and sulfur exist in ring structures, with little nitrogen existing in amine form. It is highly significant that the oxygen content is low when compared to wood. These characteristics markedly influence the pathways of combustion.

II. SOLID-PHASE PRECOMBUSTION REACTIONS FOR WOOD

Reactions of heating and drying and pyrolysis are first considered separately and then in combination.

A. Heating and Drying

Prior to pyrolysis, wood particles are heated from ambient conditions (e.g., 298 K) to that temperature where pyrolysis can occur (500-625 K). This is an endothermic process with specific heat $H_{sp(d)}$ defined by (Wenzl, 1970)

$$H_{sp(d)} = [0.064 + 0.00028(t - 273) \text{ J/g/}^\circ K \qquad (4-1)$$

where $H_{sp(d)}$ is specific heat of dry wood and t temperature (°K). The specific heat of dry wood is 1.113 J/g at 273 K (32°F) and 1.598 J/g at 373 K (212°F).

Equation (4-1) relates to dry wood. The first influence of fuel moisture content (MC) is to increase the energy required to heat the wood particle to pyrolysis temperature. With the specific heat of water being 1, and the specific heat of green wood being determined by the proportion of dry wood to water, a second expression can be constructed (Skaar, 1972):

$$H_{sp(g)} = (H_{sp(d)} + MC_{OD})/(1 + MC_{OD}) \qquad (4-2)$$

where $H_{sp(g)}$ is the specific heat of green wood, and MC_{OD} = MC, expressed on a fractional basis and calculated on an OD basis.

The second influence of moisture is to place a ceiling on the temperature of the core of the fuel particle until the water is vaporized. This influence may help define regions of reaction stages in the fuel block as the water is vaporized from the surface to the center of the particle.

The third influence of moisture is to increase the thermal conductivity of the fuel particle, thus increasing the rate of transmission of heat from the surface to the center of the particle. For wood at MC_g < 40 percent the thermal conductivity (K) equation is (U.S. Forest Products Laboratory, 1974)

$$K_t = [(5.18 + 0.096\ MC_g)S_{o,g} + 0.57FVV] \times 4.184 \times 10^{-4} \quad (4\text{-}3)$$

where K_t is given in $J/cm\text{-}°K\text{-}sec$, $S_{o,g}$ is the oven dry weight green volume specific gravity (SG), and FVV the fractional void volume. For wood where MC > 40 percent, K_t is calculated by (U.S. Forest Products Laboratory, 1974)

$$K_t = [(5.18 + 0.131MC_g)S_{o,g} + 0.57FVV] \times 4.184 \times 10^{-4} \quad (4\text{-}4)$$

From these equations one can observe that, in the initial heatup process, water increases heat conduction into fuel particles.

The heatup and drying phases are characterized principally by physical influences on moisture. These influences, however, are felt in the solid-particle pyrolysis phase.

B. Solid-Fuel Pyrolysis

In pyrolysis of the solid-fuel particle, chemical reactions dominate. Shafizadeh and Chin (1977) give pyrolysis temperature ranges as follows: hemicelluloses, 500-600 K (440-620°F); cellulose, 600-650 K (620-710°F); and lignin, 500-773 K (480-930°F). When the surface region is heated to greater than 500 K (440°F), pyrolysis begins. In pyrolysis the holocellulose and lignin fractions behave quite differently. Thus, the holocellulose and lignin fractions are treated separately here.

1. Holocellulose Pyrolysis

Holocellulose pyrolysis begins with an attack on the glyco-sidic linkages between individual glucose, xylose, arabinose, etc. units. From there proceeds a series of reactions cleaving the molecules into gaseous fragments and condensation reactions

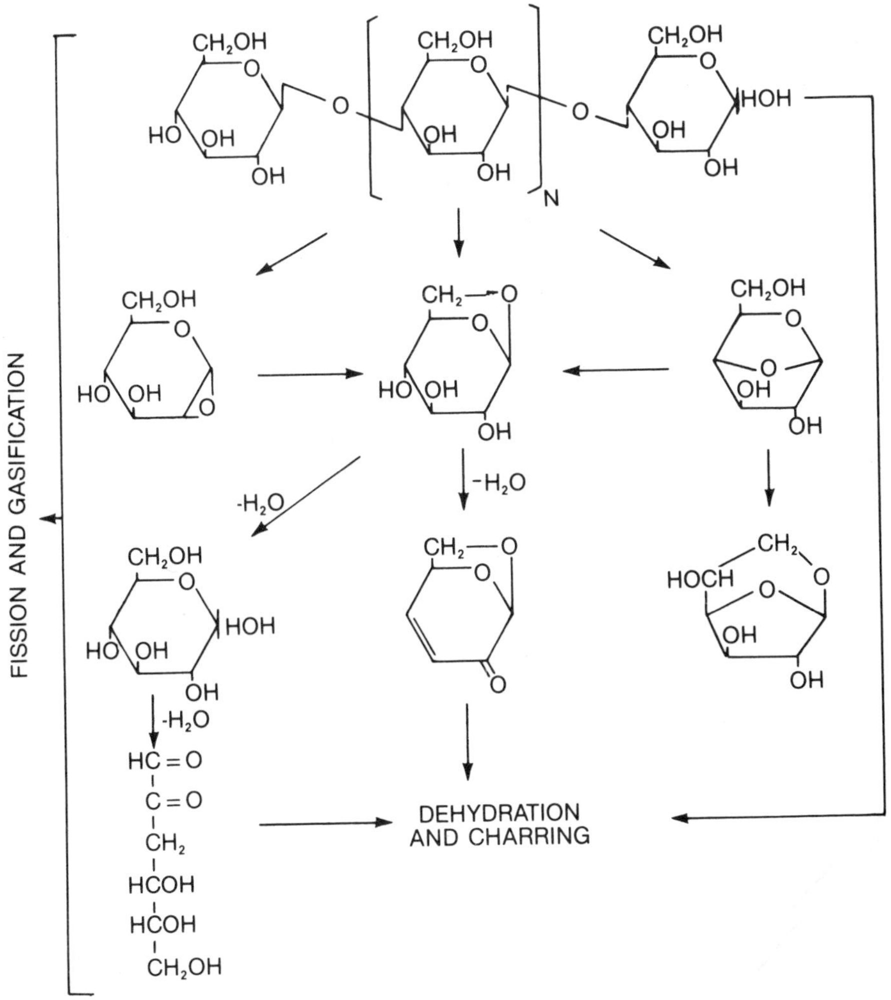

FIG. 4-3. The Shafizadeh and DeGroot (1976) mechanism for cellulose degradation by pyrolysis.

producing char. Higher temperatures favor volatile producing reactions while lower temperatures favor char production (Shafizadeh and DeGroot, 1976; Shafizadeh and Lai, 1972). Figure 4-3 shows one mechanism for cellulose thermal degradation.

The influence of ether linkages, acetyl structures, and carboxyl and carbonyl functional groups in this pyrolysis is to increase the number and complexity of the products. Figure 4-3 shows a ring opening as an intermediate step in volatile formation. Carried to its conclusion, such products as acetic acid and acetaldehyde are generated. By decarboxylation of acetic acid and decarbonylation of acetaldehye, CH_4, CO_2, and CO are produced:

$$CH_3COOH \longrightarrow CH_4 + CO_2 \tag{4-5}$$

$$CH_3CHO \longrightarrow CH_4 + CO \tag{4-6}$$

For the hemicelluloses the thermal degradation mechanisms are essentially analogous to the mechanisms for cellulose (Shafizadeh and Lai, 1975). Again, the decarboxylation and decarbonylation reactions proceed. In the latter case C_2H_4 and CO result from propenal. Table II gives products from the pyrolysis of dry cellulose and xylan at 873 and 773 K, respectively. In the combustor such products may be pyrolyzed further before oxidation.

2. Lignin Pyrolysis

The pyrolysis of lignin has been carried out both in traditional form and in helium plasma. In the former case, chars constituted 55 percent of the yield at 823 K (Allan and Mattila, 1971). In helium plasma, 33 percent of the lignin evolved into char (Graef, 1978). Gaseous products from traditional pyrolysis included CO, CO_2, CH_4, and C_2H_6 in percentages of 50, 10, 38, and 2, respectively (Allan and Mattila, 1971). The distribution of gaseous products from plasma pyrolysis was CO, 44

TABLE II. PYROLYSIS PRODUCTS OF CELLULOSE AND XYLAN[a]

Product	Cellulose (percent)[b]	Xylan (percent)[b]
Acetaldehyde	1.5	2.4
Acetone propionaldehyde	0.0	0.3
Furan	0.7	T
Propenal	0.8	0.0
Methanol	1.1	1.3
2-Methylfuran	T	0.0
2,3-Butanedione	2.0	T
1-Hydroxy-2-propanone glycoxal	2.8	0.4[c]
Acetic acid	1.0	1.5
2-Furaldehyde	1.3	4.5
5-Methyl-2-furaldehyde	0.5	0.0
Carbon dioxide	6.0	8.0
Water	11.0	7.0
Char	5.0	10.0
Tar	66.0	64.0

[a]From Shafizadeh and Chin (1977).
[b]T = trace.
[c]All 1-hydroxy-2-propane.

percent; CO_2, 2 percent; H_2, 43 percent; CH_4, 2 percent; and C_2H_2, 14 percent (Graef, 1978). Traces of C_2H_6 and higher hydrocarbons also occurred. Tars include numerous aromatic compounds including phenylacetylene, anthracene, and napthalene.

One mechanism of lignin degradation to volatiles and char was described by Connors et al. (1980). At high temperatures, benzene ring-oxygen cleavage yields aromatic rings with a resonance-stabilized radical and a methoxy radical, which may or may

TABLE III.   CHAR FORMATION AS A FUNCTION OF MATERIAL PYROLYZED[a,b]

Material	Char yield (wt. percent)
Cellulose	14.9
Poplar wood	21.7
Larch wood	26.7
Aspen (foliage)	37.8
Douglas fir bark	47.1
Klason lignin	59.0

[a]From Shafizadeh and DeGroot (1976).
[b]$T = 673$ K.

not be stabilized in the combustor by the heterogeneous gas phase reduction reactions.   Carbon-carbon cleavages also occur at the α-B, and β-5 linkages (see Fig. 4-2, structure of lignin).

C.   Combined Composition Influences

The proportion of holocellulose and lignin is one of the major determinants in the distribution of products from pyrolysis.   Shafizadeh and DeGroot (1976) demonstrated this by reporting char formation as a function of material pyrolyzed.   Their results are shown in Table III.   Shafizadeh and DeGroot attribute the influence of lignin on char formation to the presence of benzene rings.

It is useful here again to compare wood to coal.   Highly reactive lignite also yields a high volatile/fixed carbon ratio in pyrolysis; and this ratio declines to near zero as one moves through the bituminous coals to anthracite.   Simultaneously the number of benzene rings per cluster and the presence of reactive functional groups decline (Averitt, 1973; Wender, 1976).

TEMP (°K)

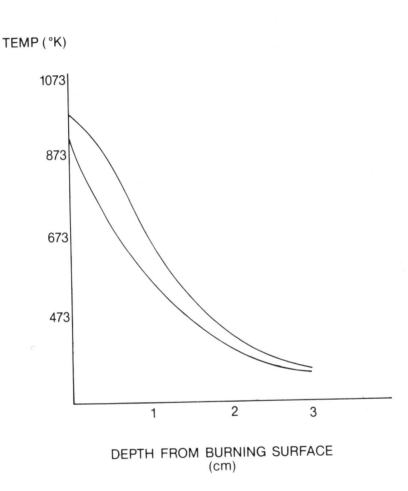

DEPTH FROM BURNING SURFACE
(cm)

FIG. 4-4. The temperature of the fuel particle as function of depth from the burning surface for Douglas fir. Note the rapid decline in temperature from approximate;y 973 to 373 K as the depth of the fuel particle approaches 2 cm (from Zicherman, 1978).

D. Influences of Moisture on Fuel Particle Pyrolysis

The effects of moisture on fuel particle pyrolysis include influencing the relative proportions of char and volatiles produced and reducing the rate of pyrolysis.

It has been shown by Wiggins and Krieger (1980) that there is a sharp physical separation between the solid-phase heating and drying zone and the pyrolysis zone. This separation largely results from the temperature gradient in the fuel particle, shown by Zicherman (1978). That temperature gradient is depicted here in Fig. 4. It is in this context that solid particle pyrolysis is examined.

It has been shown previously (Shafizadeh and DeGroot, 1976; Shafizadeh and Lai, 1972) that temperature influences the volatiles/char distribution resulting from pyrolysis. Maa and Bailie (1978) and Wiggins and Krieger (1980) also show that temperature influences the rate of pyrolysis and particle consumption. The dominant influence of moisture content is to reduce the flame temperature in the combustor, hence to drive the products toward char, and to reduce the rate of pyrolysis. Both the fuel particle surface temperature and the core temperature are reduced (see Fig. 4-4).

The influence of moisture on flame temperature can be further quantified by calculations of theoretical flame temperature. Given that the enthalpy of the reactants equals the enthalpy of the products, adiabatic flame temperature can be calculated by the following formula (Hougen et al., 1954):

$$H_p = \sum n_i \int_{298}^{T} c_{p_i} + dT + \sum n_i \lambda_i \qquad (4\text{-}7)$$

where H_p is the enthalpy of the products, n_i the moles of any product (e.g., CO_2), T (°K) the absolute temperature, c_{p_i} the molal heat capacity of any product, and λ_i the latent heat associated with any phase change.

Using this formula, the following relationship is found to exist for wood:

$$T_a = 1920 - (1.51[MC_g/(1-MC_g)] \times 100) - 5.15 Xcs_{Air} \qquad (4-8)$$

where T_a (°K) is the adiabatic flame temperature, MC_g the moisture content fraction, and Xcs_{Air} the percentage of excess air. The r^2 value for this expression is 0.89. Thus, for every oven dry percentage increase in moisture, flame temperature decreases by 1.5 K.

Beyond this direct influence on flame temperature, moisture content determines the level of excess air required according to the Villesvik spreader-stoker function (Tillman, 1980)

$$Xcs_{Air} \text{ (percent)} = 40[MC_g/(1 - MC_g)] \qquad (4-9)$$

This function holds for all fuels above 33 percent MG_g. Using Eq. (4-8), every oven dry percentage increase in moisture content decreases flame temperature by 3.6 K. Thus, fuel with 33 percent MC will burn at 1740 K, and fuel at 50 percent MC_g will burn at 1560 K. This increase in moisture content will decrease the production of volatiles from pyrolysis and increase the percentage of fuel being pyrolyzed to char (Shafizadeh and DeGroot, 1976; Shafizadeh and Lai, 1972).

Broido (1976) points to a similar influence on the volatile/char ratio stemming from ash content. The inerts act as a heat sink, reducing local temperatures and may also act as a catalyst for char formation.

In summary, solid-phase reactions of heating, drying, and pyrolysis are largely driven by fuel composition and the core's internal heating rate. Moisture affects these reactions by increasing the thermal conductivity coefficient, decreasing the flame temperature, increasing the heat capacity of the fuel particle, and thus reducing the heating rate to the point where char formation is favored.

III. GAS PHASE REACTIONS

The gas phase reactions involve a myriad of free radical pathways for complete oxidation of the volatiles formed. Three zones exist: (1) the precombustion zone, where chain initiation dominates, (2) the primary combustion zone, where chain branching is most prominent, and (3) the postcombustion zone, where chain termination reactions occur.

A. Precombustion Reactions

Precombustion reactions involve homolytic cleavage of the volatiles into radical fragments by the two generic pathways

$$R - R \longrightarrow R + R' \qquad\qquad\qquad (4\text{-}10)$$

$$R'' - H \longrightarrow R'' + H \qquad\qquad\qquad (4\text{-}11)$$

For reaction (4-10) applied to C_2H_6, activation energies are typically 368 kJ/mole. Reaction (4-11) has an activation energy of 410 kJ/mole (Edwards, 1974). Because E_{act} is higher with reaction (4-11) than (4-10), reaction (4-11) is considered to hold only when R'' represents a methyl group. If R'' contains two or more carbons, a C-C bond will be broken rather than a C-H bond. Radicals such as CH_3 may also emerge directly from solid-fuel pyrolysis such as lignin pyrolysis (Connors et al., 1980).

Edwards (1974) shows details associated with the pathway identified by reaction (4-10) as follows:

$$C_2H_6 + M \longrightarrow 2CH_3 + M \qquad\qquad\qquad (4\text{-}12)$$

$$2CH_3 + 2C_2H_6 \longrightarrow 2CH_4 + 2C_2H_5 \qquad\qquad\qquad (4\text{-}13)$$

$$M + C_2H_5 \longrightarrow H + C_2H_4 + M \qquad\qquad\qquad (4\text{-}14)$$

$$H + C_2H_6 \longrightarrow H_2 + C_2H_5 \qquad\qquad\qquad (4\text{-}15)$$

The species M is any heat-removing molecule or particle (e.g, ash). The pyrolysis of gaseous molecules in the gas phase reaction zone is more likely to produce sequence 12 to 15 than to dissociate H_2O into H and OH, as that reaction has E_{act} = 498 kJ/mole. Some dissociation does occur, however, increasing the OH concentration.

The passing of steam through this zone also removes energy that would otherwise be available for radical generation. Edwards (1974) observed that ash passing through the reaction zone, like water, serves as a heat sink. Ash also reradiates energy to facilitate pyrolysis and the all-important radical formation processes in this oxygen-deficient region.

In addition to chain initiation reactions, the precombustion gas phase zone incudes reduction reactions. Radicals may collide and recombine as follows:

$$R + R' \longrightarrow R - R' \qquad (4-16)$$

These reactions are particularly prevalent if the precombustion zone is spatially large. Such a large zone can be accomplished by use of substoichiometric quantities of primary (undergrate) air and a relatively high placement of overfire air tuyeres in the firebox. In such cases, the flaming combustion is spatially separated from the char bed on the grate.

Heterogeneous reduction reactions are quite useful in controlling the formation of NO_x from fuel-bound nitrogen. The nitrogen based radicals recombine and the fuel nitrogen is given off as N_2. This phenomenon is discussed more completely in Chapter 6.

B. Primary Combustion Reactions

Oxygen and fuel, when mixed in the primary combustion zone, undergo a series of free radical reactions ultimately producing CO_2 and H_2O. Reactions of the form $RH + O_2 \longrightarrow R + HOO$

are not energetically favored (Edwards, 1974). Rather, the radicals formed in the pyrolysis zone and the precombustion zone are the probable initiators.

Hay (1974) shows one highly probable mechanism for chain branching formation of the key reactive species:

$$CH_3 + O_2 + M \longrightarrow CH_3O_2 + M \tag{4-17}$$

$$CH_3O_2 \longrightarrow CH_2O + OH \tag{4-18}$$

This sequence of reactions not only generates hydroxy radicals but also the key combustion intermediate CH_2O. From CH_2O, HCO and then CO are formed by reaction with OH. CH_2O is further the key intermediate resulting from CH_4 oxidation via OH and O attack. The concentration of CH_2O reaches its maximum in flames at 1320 K, well in the temperature range of wood combustion (Palmer, 1974).

C. Postcombustion Reactions

For relatively low temperature processes (e.g., 1270 to 1970 K) such as wood combustion, chain termination reactions occur in the secondary reaction region (Edwards, 1974). Again, the hydroxy radical plays a significant role when present in relatively large concentrations. Palmer (1974) argues that of the following, reactions (4-19) and (4-20) are relatively fast compared to reaction (4-21), and thus reaction (4-20) is of minor importance:

$$HCO + OH \longrightarrow CO + H_2O \tag{4-19}$$

$$CO + OH \longrightarrow CO_2 + H \tag{4-20}$$

$$CO + O_2 \longrightarrow CO_2 + O \tag{4-21}$$

Thus he states that formation of CO_2, particularly from CO, is controlled by the OH concentration, which he has shown to be relatively high in lower-temperature systems. Concluding the chain termination is the recombination of radicals such as H and OH aided by a heat-removing species (M).

In gas phase combustion, it is significant to note the high molar ratio C:H in the initial wood feedstock vis-a-vis coal, which is 1:1.45 for softwoods and 1:1.37 for hardwoods (Antal, 1979). As the discussion of solid phase pyrolysis showed the hydrogen yields not only pyrolysis water but also significant quantities of CH_4, C_2H_4, and C_2H_6. Thus, substantial quantities of hydrogen exist in the volatile gases to increase hydroxy radical formation for complete and rapid oxidation. While equilibrium and rate constants for many of these individual reactions exist (Jensen and Jones, 1978), no overall expressions are in the literature due to the numerous variables associated with oxidation of volatiles from wood.

IV. CHAR COMBUSTION

The char evolved by pyrolysis is quite porous and contains numerous free-radical sites for O_2 attack (Bradbury and Shafizadeh, 1980). Further, it contains some hydrogen and oxygen. Pober and Bauer (1977) found up to 4 percent hydrogen (weight basis) in the pyrolysis chars of wood bark. Bradbury and Shafizadeh (1980) describe an empirical formula for cellulose char of $C_{6.7}H_{3.3}O$.

Numerous mechanisms have been proposed to describe char oxidation, recognizing that the rate of char combustion is limited by the number of free radical sites on the porous char surface (Wallouch and Heintz, 1974). Char oxidation is also considered to be mass transport limited (Glassman, 1977).

The classic Boudouard reaction is proposed by Glassman (1977) as a general initiator of char combustion:

$$C + CO_2 \longrightarrow 2CO \qquad\qquad (4\text{-}22)$$

This reaction is highly endothermic and has an equilibrium constant of 1.1×10^{-2} at 800 K and 1.9 at 1100 K (Antal, 1979). At 1200 K the K_{eq} value is 57.1 (Antal, 1979). Under this scheme the CO is released in volatile form and oxidation is completed in the flaming combustion.

The research of Bradbury and Shafizadeh (1980) demonstrates that oxygen may be chemisorbed directly onto the char. Chemisorption of O_2 on the porous char surface has an activation energy of 54 kJ/mole to 105 kJ/ mole and increases linearly as the quantity of O_2 chemisorbed increases from 0 to 2.5 mole O_2/g char (Bradbury and Shafizadeh, 1980). Based on this research Bradbury and Shafizadeh propose reactions the following reactions as one principal char oxidation process:

$$C* + O_2 \longrightarrow C(O)* \longrightarrow C(O)_m \xrightarrow{\text{fast}} CO + CO_2 \qquad (4-23)$$

$$C* + O_2 \longrightarrow C(O)_s \xrightarrow{\text{slow}} CO + CO_2 \qquad (4-24)$$

The asterisk indicates an active site for the reaction, the subscript m a mobile species, and the subscript s a stable specie. These active sites could be generated by such pyrolysis mechanisms as the one proposed by Connors et al. (1980).

A third mechanism for char oxidation is proposed by Mulcahy and Yang (1975). It involves reaction of hydroxy radicals with active sites on the char as follows:

$$2OH + C \longrightarrow CO + H_2O \qquad (4-25)$$

$$OH + CO \longrightarrow CO + H \qquad (4-26)$$

Hydroxy radicals necessary for reaction (4-25) to proceed may be generated internally by homolytic cleavage of the many hydroxy functional groups in the wood. They may come from dissociation of the moisture being released from the fuel. In all probability all of the mechanisms discussed above play roles in char combustion.

The role of moisture in char oxidation is less defined than the role of moisture in solid particle pyrolysis. It may provide hydroxy radicals, as previously discussed. Pompe and Vines (1966) speculate that moisture "smothers" reaction sites, reducing the rate of char oxidation. Certainly by decreasing the flame temperature in the combustor, the presence of moisture retards the rate of char oxidation.

V. SUMMARY

The Edwards conceptual model for solid fuel combustion facilitates consolidation of much research with wood combustion. Through this mechanism much of the complexity associated with wood combustion can be elucidated. Further, the multifaceted role of moisture in wood combustion can be examined.

Wood combustion is a multistaged process involving heating and drying, solid-particle pyrolysis to volatiles and char, gas phase volatile oxidation, and char oxidation. The many functional groups in the wood constituents contribute to the plethora of solid particle pyrolysis products shown in Table II. The many functional groups, oxygen content, and domination of aliphatic structures contribute to the increased reactivity of wood relative to coal. Specifically, these characteristics contribute to the high proportion of flaming combustion relative to char oxidation for wood, when compared to the higher-ranked coals.

The moisture content of wood increases the energy required to heat the fuel particle to pyrolysis temperature, increases the thermal conductivity of the fuel particle, and directs the solid-particle pyrolysis process toward increasing char production. It may contribute to increasing concentration of hydroxy radicals for gas phase and char reactions. It may reduce the rate of char oxidation by decreasing the temperature of char ox-

idation and by smothering reactive sites. Like the process of
wood combustion itself, the role of moisture in the process is
highly complex, as a review of combustion mechanisms shows.

Based upon this description of the combustion process, the
generation of useful and economically valuable heat can be understood. Further, the formation of particulates and NO_x,
effluents that must be controlled, can be discussed. In particular, this process description provides critical insights into
the behavior of specific systems for wood firing.

Given this process description, Chapter 5 presents the production of useful heat in terms of combustion efficiency, flame
temperature, and rates of heat release. Chapter 6 considers the
formation and control of airborne emissions.

REFERENCES

Allan, G., and Mattila, J. (1971). High energy degradation,
in "Lignins: Occurrence, Formation, Structure, and Reactions"
(K. Sarkanen and H. Ludwig, eds.), pp. 575-592. Wiley (Interscience), New York.

Antal, M. J. (1979). "Thermochemical Conversion of Biomass:
The Scientific Aspects." A Report to the Office of Technology
Assessment of the Congress of the United States. Princeton Uni-
versity.

Arola, R. (1976). Wood fuels--How do they stack upµ in
"Energy and the Wood Products Industry." Proc. Forest Prod.
Res. Soc., 34-41. Madison, Wisconsin.

Averitt, P. (1973). Coal, in "United States Mineral Resources," Geological Survey Professional Paper 820 (D. Brobst
and W. Pratt, eds.). USGPO, Washington, D.C.

Bailie, R. (1977). Current developments and problems in
biomass gasification, Proc. Forest Fuels Symp., Winnipeg, Canada.

Bradbury, A. G. W., and Shafizadeh, F. (1980). Role of oxygen chemisorption in low-temperature ignition of cellulose, Combustion Flame 37:85-89.

Broido, A. (1976). Kinetics of solid-phase cellulose pyrolysis, in "Thermal Uses and Properties of Carbohydrates and Lignins (F. Shafizadeh, K. V. Sarkanen, and D. A. Tillman, eds.), pp. 19-36. Academic Press, New York.

Connors, W. J., Johanson, L. N., Sarkanen, K. V., and Winslow, P. (1980). Thermal degradation of kraft lignin in tetralin, Holzforschung 34:29-37.

Cowling, E. B., and Kirk, T. K. (1976). Properties of cellulose and lignocellulosic materials as substrates for enzymatic conversion processes, in "Enzymatic Conversion of Cellulosic Materials: Technology and Applications" (E. G. Gaden, M. H. Maudels, E. T. Reese, and L. A. Spano, eds.), pp. 95-124. Wiley (Interscience), New York.

Edwards, J. (1974). "Combustion: Formation snd Emission of Trace Species." Ann Arbor Science Publ., Ann Arbor, Michigan.

Glassman, I. (1977). "Combustion." Academic Press, New York.

Graef, M. (1978). "Reactions of Lignin in Microwave Induced Plasma." M.Sc. Thesis, Univ. of Washington, Seattle.

Hay, J. M. (1974). "Reactive Free Radicals." Academic Press, New York.

Hougen, O. F., Watson, K. A., and Razatz, R. A. (1954). "Chemical Process Principles, Part 1: Material and Energy Balances," 2nd ed. Wiley, New York.

Jensen, D. E., and Jones, G. M. (1978). Reaction rate coefficients for flame calculations, Combustion Flame 32:1-34.

Junge, D. C. (1975). "Boilers Fired with Wood and Bark Residues." Res. Bull. 17, Forest Research Laboratory, Oregon State Univ., Corvallis.

Maa, P. S., and Bailie, R. C. (1978). Experimental pyrolysis of cellulosic material, Proc. 1978 AICHE 84th Natl. Mtg., Atlanta, Georgia, February 26-March 1.

Mulcahy, M. F. R., and Yang, C. C. (1975). The reaction of hydroxyl radicals with carbon at 298 K, Carbon 13:115-124.

Palmer, H. B. (1974). Equilibria and chemical kinetics in flames, in "Combustion Technology: Some Modern Developments" (H. B. Palmer and J. M. Beer, eds.), pp. 2-33. Academic Press, New York.

Pober, K. W., and Bauer, H. F. (1977). The nature of pyrolytic oil, in "Fuels from Waste" (L. L. Anderson and D. A. Tillman, eds.), pp. 73-86. Academic Press, New York.

Pompe, A., and Vines, R. G. (1966). The influence of moisture on the combustion of leaves, Austral. Forestry 231-241.

Sarkanen, K. V. (1971). Precursors and their polymerization, in "Lignins: Occurrence, Formation, Structure, and Reactions" (K. Sarkanen and H. Ludwig, eds.), pp. 95-164. Wiley (Interscience), New York.

Shafizadeh, F., and Chin, P. S. (1977). Thermal deterioration of wood, in "Wood Technology: Chemical Aspects" (I. S. Goldstine, ed.), pp. 57-81. ACS Press, Washington, D.C.

Shafizadeh, F., and DeGroot, W. F. (1976). Combustion characteristics of cellulosic fuels, in "Thermal Uses and Properties of Carbohydrates and Lignins (F. Shafizadeh, K. V. Sarkanen, and D. A. Tillman, eds.). Academic Press, New York.

Shafizadeh, F., and DeGroot, W. F. (1977). Thermal analysis of forest fuels, in "Fuels and Energy from Renewable Resources" (D. A. Tillman, K. V. Sarkanen, and L. L. Anderson, eds.), pp. 93-114. Academic Press, New York.

Shafizadeh, F., and Lai, Y. Z. (1972). Thermal degradation of 1-6-anhydro-β-D-glucopyranuse, J. Org. Chem. 37:278-284.

Shafizadeh, F., and Lai, Y. Z. (1975). Thermal degradation of 2-deoxy-D-arabino-hexonic acid and 3-deoxy-D-ribohexono-1,4-kutone, Carbohydrate Res. 42:39-53.

Skaar, C. (1972). "Water in Wood." Syracuse Wood Science Series No. 4, Syracuse Univ. Press, Syracuse, New York.

Sohn, H. Y., and Szekely, J. (1972). A structural model for gas-solid reactions with a moving boundary, III, Chem. Eng. Sci. 27:763-776.

Tillman, D. A. (1980). Fuels from waste, in "Kirk-Othmer Encyclopedia of Chemical Technology," Vol. 2, 3rd ed. Wiley, New York.

U.S. Forest Products Laboratory (1974). "Wood Handbook: Wood as an Engineering Material." USGPO, Washington, D.C.

Wallouch, R. W., and Heintz, E. A. (1974). The oxidation of graphite: Zone II omission, Carbon 12:243-248.

Wender, I. (1976). Catalytic sybthesis of chemicals from coal, Catal. Rev. 14:97-129.

Wenzl, H. (1970). "The Chemical Technology of Wood." Academic Press, New York.

Wiggins, D., and Krieger, B. B. (1980). Effect of heating and quenching rates on volatiles produced from combustion-level-heat-flux pyrolysis of biomass, Proc. AICHE 89th Natl. Mtg., Portland Oregon.

Zicherman, J. (1978). A study of wood morphology and microstructure in relation to its behavior in fire exposure. Ph.D. Thesis, University of California, Berkeley.

CHAPTER 5

HEAT PRODUCTION AND RELEASE FROM WOOD COMBUSTION

I. INTRODUCTION

Typically, the reactions of combustion are expressed by

$$C + O_2 \longrightarrow CO_2 + 393 \times 10^3 \quad J/g\text{-mol } C \qquad (5\text{-}1)$$

$$H_2 + 0.5O_2 \longrightarrow H_2O + 266 \times 10^3 \quad J/g\text{-mol } H_2 \qquad (5\text{-}2)$$

Complete oxidation yields CO_2, H_2O, and heat. A more complex treatment of the products of wood combustion involves examining the quantity of useful heat released, the flame temperatures achieved, and the rates of heat release.

The production of useful heat is influenced not only by fuel conditions and the process of wood combustion, but also by the method of wood combustion. Firing systems discussed here include pile burners, spreader stokers, suspension fired systems, and fluidized bed combustors. Emphasis, however, is on the spreader stoker as it is the most prominent industrial system.

A. Pile-Burning Systems

The traditional Dutch oven, the Wellons cell burner, and the inclined grate are all examples of pile-burning systems. Of these, the inclined grate, shown in Fig. 5-1, is discussed here. The inclined grate is considered due to its size flexibility. Systems have been built ranging from <25 GJ/h to >400 GJ/h (Envirosphere 1980; MacCallum 1979).

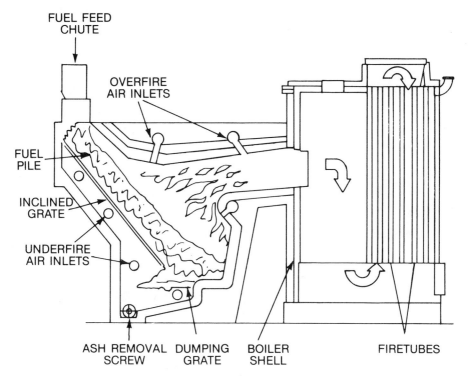

FIG. 5-1. An inclined-grate combustor of wood fuel. In this system, heating and drying can occur relatively close to the fuel feed shoot. Solid-phase pyrolysis can occur as the fuel is sliding down the grate. Char oxidation can occur at the base of the grate and on the dumping grate. Gas phase reactions can be controlled by overfire air distribution and separated completely from solid phase reactions (from Envirosphere, 1980).

In the inclined-grate system, fuel is fed to the top of the grate, where heating and drying occurs. As the fuel slides down the grate, solid-phase pyrolysis reactions occur and the volatiles evolved enter a secondary combustion chamber. The char moves further down the grate, where char oxidation takes place. By controlling the (primary) air supply to the char, volatile oxidation near the higher portions of the grate is avoided. Finally, the ash is removed from the toe of the grate. Volatiles evolved undergo precombustion reactions prior to being contacted by the secondary air stream used to support flaming oxidation. Postcombustion gas phase reactions occur in the large boiler area.

It is significant that the sloping-grate system allows virtually complete separation of combustion reaction zones. Control of primary air for char oxidation and secondary air for volatile oxidation ensures the separation of combustion reaction zones. Separation is sufficiently complete that some inclined-grate systems are considered close-coupled gasifiers.

B. Spreader-Stoker Systems

The spreader-stoker is the workhorse of industrial wood burning equipment. Figure 5-2 schematically shows this design. In the spreader-stoker, fuel particles are fed into the firebox and flung, mechanically or pneumatically, across the grate. Some heating and drying, and possibly some pyrolysis, occurs while the particle is in suspension. For the most part, however, solid-phase pyrolysis and char oxidation occur on the grate. Precombustion gas phase reactions occur between the grate and the zone where secondary air is introduced. Gas phase oxidation occurs either throughout the firebox if most air enters as primary air, or in the vicinity of the zone where secondary air is introduced if the undergrate air is limited to substoichiometric quantities.

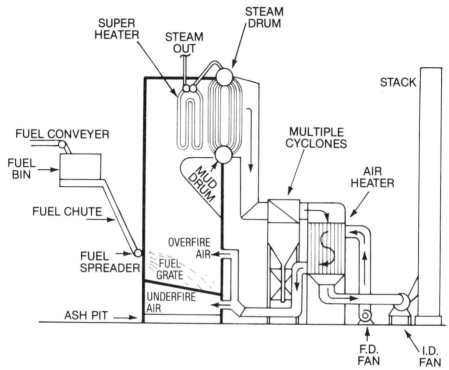

FIG. 5-2. The spreader-stoker system including steam drum, superheater, air pollution control equipment, and air heater. Note, with respect to the spreader-stoker, the partial pyrolysis and combustion of fuel above the grate, with final solid-fuel reactions occurring on the grate. Note also the ability to separate gas phase reactions from gas-solid reactions by placement of the overfire air port, and by distrubution of the air (from Junge, 1975).

As the above discussion shows, the distribution of air governs the degree to which combustion reactions can be physically separated in spreader-stokers. If the sharp distribution of air recommended by Junge (1979) is used, separation of reaction zones is accomplished. Further, a substantial precombustion gas phase reaction zone can be created in order to promote reducing and radical recombination reactions.

Like pile burning, spreader-stoker firing involves grate firing, and the grate design options are varied. Fixed or moving grates may be installed. Fixed grates may be constructed of alloy steels or refractory materials. They may be cooled by use of ambient primary air or water. Moving grates are made from high-temperature alloy steels (Junge, 1979). Grate materials are particularly important as temperatures of 1900 K (3000°F) have been achieved on the grate in wood combustion (Kester, 1980).

C. Suspension-Burning Systems

Suspension burning involves firing finely divided (e.g., <2 mm) dry particles (<15 percent MC) in a primary air stream. All combustion reactions occur while the particle is in midair. The combustion reactions in each particle occur sequentially; that is, heat-up occurs followed by pyrolysis, gas phase reactions, and carbon oxidation.

Because a large air stream is required to keep the fuel particles in suspension, physical separation of reaction zones is difficult to accomplish, but staging is possible in terms of a partial separation of precombustion gas phase reactions and oxidation reactions through the use and control of a secondary air stream. Suspension firing is essentially analogous to oil burning in the creation of diffusion flames.

D. Fluidized-Bed Combustion Systems

Fluidized-bed combustion involves supporting the fuel in a partially suspended bed of inert material such as limestone or sand. Primary or undergrate air supports the bed while providing the oxygen for combustion reactions.

In fluidized-bed combustion, all combustion reactions take place in the same zone. Further, the motion of the bed vastly improves heat transfer into the fuel particle and continuously removes the char from the fuel particle, exposing fresh fuel for reactions. Mechanistically, it is virtually impossible to physically separate reaction zones in fluidized-bed combustion.

The volumes of excess air used in fluidized-bed combustion are typically 100-140 percent. This maintains the combustion temperature below the point of ash softening or ash fusion. It is absolutely essential to hold the combustion temperature below ash fusion due to the large mass of inert bed material. This limits the efficiency of the system.

E. Firing Methods and Production of Useful Heat

System efficiency and flame temperature can be considered without regard to firing method. However, systems such as suspension burning require dry fuels. Thus, more efficient combustion may appear to be a built-in advantage. Both suspension burning and fluidized-bed combustion normally require 100 percent excess air or more, however, in order to maintain combustion below ash fusion temperature. Hence, efficiency and flame temperature are, in fact, sacrificed.

Rates of heat release are largely a function of firing method. Again, the reactivity of the fuel plays a substantial role. However, that can be considered largely as a constant, with the moisture content being the major variable. Thus, in this discussion, rates of heat release are more related to firing method than efficiency or flame temperature.

II. THE PRODUCTION OF USEFUL HEAT

Useful heat is the derived product of wood combustion. Two measures are considered here: thermal efficiency and flame temperature. The first measures the quantity of heat available for processes. The second measures the quality of heat available, and provides a thermodynamic efficiency limit to any given process.

A. The Efficiency of Wood Combustion

The following basic parameters govern the production of heat available for use: (1) enthalpy of the fuel, (2) moisture content of the fuel, (3) level of excess air employed, and (4) final stack temperature as influenced by the degree of air and/or feedwater preheating.

There are several methods for approximating thermal efficiency including the Ince (1977) formulas and several linear regression techniques (see Tillman, 1978). More precise calculations are presented here using the Villesvik excess air function for spreader stokers shown in Chapter 4 [Eq. (4-9)] and assuming a stack temperature of 450 K (350°F). The wood assumed is the softwood reported in Chapter 4.

For efficiency calculations two basic formulas are used: useful heat available,

$$H_u = HHV - HL \tag{5-3}$$

and thermal efficiency,

$$\eta = [1 - (HL/HHV)] \times 100 \tag{5-4}$$

where H_u is useful heat released, HHV the higher heating value of the fuel, and HL represents the heat losses from stack gas, ash (including unburned carbon), and radiation and other losses.

If one assumes 98-99 percent carbon conversion efficiency (Envirosphere 1980), unburned carbon losses are on the order of 30-40 kJ/100 g (14-28 x 10^3 Btu/lb) fuel burned. Radiation and other losses can be taken at 4-5 percent of the higher heating value. Thus, the critical losses are in the stack gas,

TABLE I.   MEAN MOLAL HEAT CAPACITIES (J/g-mole-°K) FOR SELECTED STACK GASES BETWEEN 298 AND t (°K)[a]

			Gas		
t (°K)	N_2	CO	O_2	H_2O	CO_2
373	29.0	29.2	29.6	33.8	38.7
473	29.1	29.4	30.0	34.2	40.6
573	29.2	29.6	30.5	34.4	42.3
673	29.2	29.9	31.0	35.2	43.8
773	29.3	30.2	31.4	35.7	45.1

[a]From Hougen et al. (1954).

and are calculated as follows:

$$HL_{sg} = \sum_{i=1}^{n} m_i(C_{p_i} \Delta T) + m_{H_2O}\lambda_{H_2O} \qquad (5\text{-}5)$$

where i designates individual stack gases (e.g., N_2, O_2, CO_2, H_2O), m the moles of any stack gas produced, C_{pi} the mean molal heat capacity between final stack temperature and ambient conditions, ΔT the temperature differential between the stack and ambient conditions, m_{H2O} indicates the moles of H_2O, and λ_{H2O} the heat of vaporization for water. Selected values for C_{pi} are presented in Table I. These values are from Hougen et al. (1954) and are for the temperature range 373-773 K.

In order to show the thermal efficiencies, combustion of 100 g dry wood with two moisture contents is assumed: 17 and 50 percent (green basis). The empirical formulas of the fuel are

$$\text{dry wood} = H_6C_{4.4}O_{2.4}N_{0.02}(H_2O)_{1.1} \qquad (5\text{-}6)$$

$$\text{wet wood} = H_6C_{4.4}O_{2.4}N_{0.02}(H_2O)_{5.6} \qquad (5\text{-}7)$$

It is useful to observe here that the total quantity of dry fuel is 120 g, and the total quantity of wet fuel is 200 g. However, both fuels have the same total heat content or the same dry wood content.

Given that the dry wood is burned with 25 percent excess air and the wet wood is burned with 40 percent excess air, the appropriate combustion equations (assuming 100 percent carbon conversion to CO_2) are as follows:

Dry wood combustion:

$$H_6C_{4.4}O_{2.4}N_{0.01}(H_2O)_{1.1} + 4.4O_2 + 16.6N_2$$
$$\longrightarrow 4.1H_2O + 4.4CO_2 + 0.9O_2 + 16.6N_2 \tag{5-8}$$

Wet wood combustion:

$$H_6C_{4.4}O_{2.4}N_{0.01}(H_2O)_{5.6} + 4.9O_2 + 18.4N_2$$
$$\longrightarrow 8.6H_2O + 4.4CO_2 + 1.4O_2 + 18.4N_2 \tag{5-9}$$

It is useful to examine these equations and note that (1) the moisture content of the stack gas more than doubles as the moisture content goes from 17 to 50 percent, and (2) the volume of stack gas increases from 26 to 33 moles. The fuel nitrogen, in effect, is disregarded in this calculation.

Table II present the calculated heat losses for these cases assuming a stack temperature of 450 K (350°F), and the combustion of 100 g (0.22 lb) of dry fuel. For an increase in moisture of 80 g (0.18 lb), the losses increase by 117 kJ (111 Btu). One can assume a fuel heat content of 19.8 kJ/g, (8500 Btu/lb), and that losses due to radiation are 4 percent and other losses due to ash and unburned carbon are about 2 percent. The combustion efficiency of the 17 percent moisture fuel is ~79 percent and the combustion efficiency of the 50 percent moisture fuel is ~68 percent. That is, combustion of 120 g (0.26 lb) of 17 percent MC fuel will release 1.47 MJ (1.31 x 10^3 Btu) of useful heat energy, while the combustion of 200 g (0.44 lb) of 50 percent MC fuel will release 1.35 MJ (1.38 x

TABLE II. CALCULATED HEAT LOSS IN STACK GASES
FOR WOOD COMBUSTION[a,b]

Stack gas component	Mean molal heat capacity (T = 298-450 K)	Fuel case, percent MC	
		17[c]	50[d]
CO_2	40.6	27,000	27,000
O_2	30.9	4,200	6,500
N_2	29.5	74,000	82,000
H_2O	35.0	21,700	45,500
H_2O heat of vaporization	40,656	167,000	350,000
Total	--	294,000	511,000

[a]From Houge net al. (1954).
[b]Values in joules.
[c]Assumes 120 g wood fuel (100 g dry wood).
[d]Assumes 200 g wood fuel (100 g dry wood).

10^3 Btu) of useful heat energy. These are summarized in Fig. 5-3.

The influence of increasing the levels of excess air can be seen from the O_2 and N_2 values in Table II. Increasing the excess air level by 15 percent caused an additional 10.3 kJ of heat loss/100 g dry fuel burned--or about 0.5 percent of the higher heating value of the fuel. This assumes dry air. On humid days, these losses become more severe due to moisture content in the combustion air.

The influence of increasing the stack temperature stems from increasing the ΔT values between stack gas and ambient conditions. It also stems from increasing the molal heat capacities of the stack gas as shown in Table II.

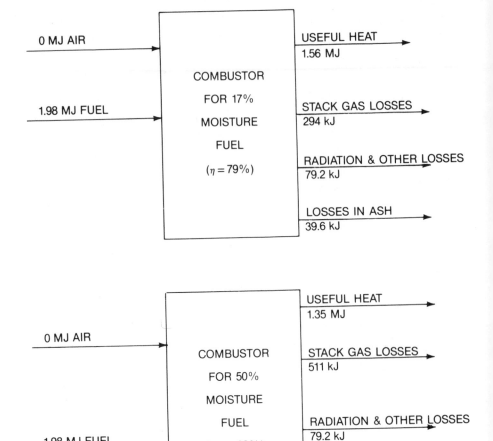

FIG. 5-3. Calculated heat balances for combustors using 17 and 50 percent MC fuel, holding energy input to the combustor constant. With respect to these two systems, it is useful to note that the quantity of fuel must be increased as moisture content is increased, and that stack gas losses are significantly increased as a function of moisture entering with the fuel.

Using the values in Table I, one can recalculate the efficiency of the 17 and 50 percent MC combustion cases. Assuming a final stack temperature of 573 K (572°F), the stack losses are 399 kJ in the combustion of 120 g of 17 percent MC fuel (25 percent excess air), and 805 kJ in the combustion of 200 g of 50 percent MC fuel (40 percent excess air). Final combustion efficiency values are 74 and 53 percent, respectively.

In summary, for a variety of reasons 450 K (350°F) is considered to be a reasonably attainable optimum stack temperature. Thus, the 79 and 68 percent efficiencies are considered to near the practical limits of thermal efficiency depending upon moisture content.

B. Flame Temperatures in Wood Combustion

Flame temperature determines the quality of heat available from combustion and establishes the maximum attainable thermal efficiency of converting the useful heat into work (e.g., generating electricity). Chapter 4 presented a simple method for flame temperature approximation [see Eq. (4-7)]. However, more precise calculations are available by using the NASA combustion model.

The NASA combustion model calculates flame temperature based upon equilibrium constants for a wide array of combustion reactions, and accounting for dissociation of final products such as CO_2. It can be used to calculate not only flame temperature but also final combustion products assuming complete combustion and then achieving equilibrium on all reversible reactions.

L. L. Anderson (personal communication of computer results, March 21, 1980) applied this model to wood fuel, assuming moisture contents of 20, 33, and 50 percent, and excess air levels up to 100 percent levels. No air preheat was assumed. Five species of wood were employed. The values resulting from poplar combustion are shown in Fig. 5-4. It is significant to note the

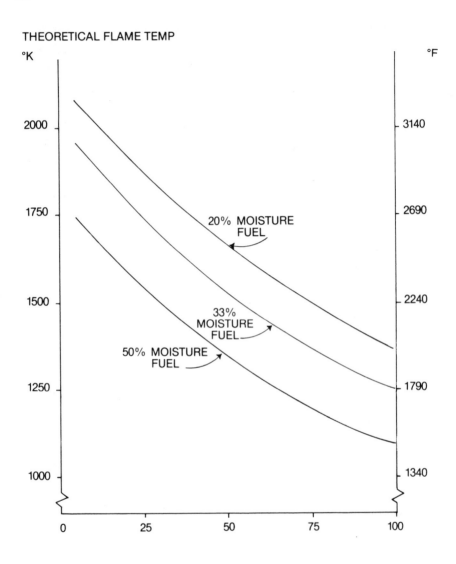

THEORETICAL FLAME TEMP

PERCENT EXCESS AIR

FIG. 5-4. Theoretical flame temperature as a function of fuel moisture content and percent excess air for Douglas fir bark. This illustrates the influence of both moisture content and excess air on flame temperature and demonstrates that excess air plays a more significant role in reducing flame temperature than does moisture content (courtesy of L. L. Anderson).

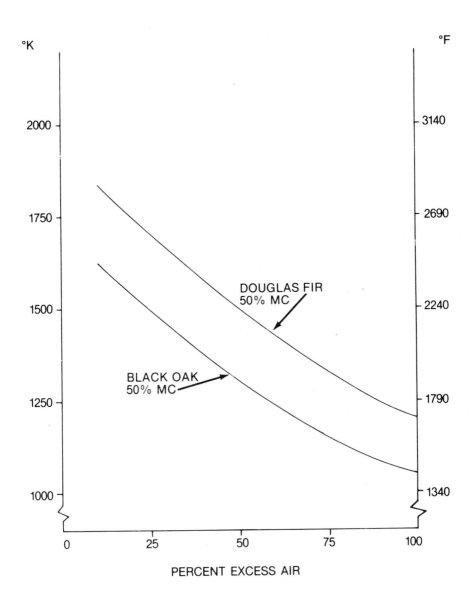

FIG. 5-5. Theoretical flame temperature as a function of percent excess air for 50 percent moisture content fuels. Note that differences in species can influence flame temperature by approximately 200 K (courtesy of L. L. Anderson).

serious influence of excess air on flame temperature, as previously documented. Figure 5-5 shows the range of values obtained for 50 percent MC fuel assuming Douglas fir bark at the high end of the range and black oak at the low end of the range.

Preheating air can be used for several purposes, including fuel drying and flame temperature increase. Typical combustion air temperatures, in systems with air heaters, range from 420 to 560 K (300-500°F).

The data in Figs. 5-4 and 5-5 permit drawing certain conclusions: (1) in order to achieve high flame temperature, it is most important to control excess air levels; (2) moisture is more important in terms of controlling excess air levels than influencing flame temperature directly. Preheating air may be more important to overall system (first law) efficiency than increasing the flame temperature.

It should also be noted that the metallurgical limit of steel used in boilers is ~840 K (1050°F) within current practice (Hill, 1977). Properly fired wood boilers achieve flame temperatures well above that level. Thus, combustion temperature does not limit raising high-pressure/high-temperature steam.

C. Relation of Heat Production to Firing Methods

Thermal efficiency and flame temperature may be considered independently of firing method. However, given that wood ash fusion occurs in the 1420-1650 K (2100-2500°F) range and that fluid bed ash fusion can occur as low as 1230 K (1750°F), the control of excess air to minimize clinker formation may be necessary. In such cases, combustion efficiency may have to be sacrificed. For suspension fired equipment and fluidized-bed systems, excess air levels of 100 percent are common. Consequently, the production of useful heat, particularly at high temperatures, is reduced significantly. Pile-burning and spreader-stoker systems can be more efficient. Undergrate (primary) air may remain at ambient temperatures (e.g., 298 K) to reduce clinker formation, while overfire air may contain the heat recovered from the stack gas in air preheaters (Junge, 1979). High temperature and overall combustion efficiencies in the 68-79 percent range can be maintained.

D. Comparison of Wood and Coal

It is useful to compare thermal efficiencies and flame temperature between wood and coal. This is particularly so in light of boiler heat transfer coefficients. Typical values are 65 kJ/m^2-°K-h (10 Btu/ft^2-°F-h) for spreader-stokers. For fluidized-bed combustors, values are 325 KJ/m^2-°K-h (50 Btu/ft^2-°F-h). Lower flame temperatures increase heat transfer surface requirements.

For bituminous coal fired units, typical thermal efficiencies are 80-85 percent. Moisture, so prominent in wood combustion, plays a very small part. Similarly, theoretical flame temperatures are slightly higher, with typical values being 2030-2250 K (3200-3600°F). Theoretical flame temperatures can exceed 2480 K (4000° F). It should be noted that levels of excess air are in the 6-15 percent range, however (Fryling, 1966).

III. RATES OF HEAT RELEASE

While the combustion of wood is less efficient than the combustion of coal, rates of heat release are substantially higher. This results from the reactivity related to the higher volatile content of wood fuel relative to coal. Proximate analyses of a wide range of wood fuels show an approximate volatile:fixed carbon ratio of 3:1 for most wood fuels (Arola, 1976). This contrasts with coal, where ratios may be <1:1 (Anderson, 1977). Wood is a more reactive fuel.

TABLE III. FIRING METHODS AND RATES OF HEAT RELEASE[a,b]

| | Heat release rate | |
Firing method	SI units	English units
Pile burning	8.5 GJ/m^2	725 X 10^3 Btu/ft^2
Inclined grate	3.5 GJ/m^2	300 x 10^3 Btu/ft^2
Spreader-stoker	10.4 GJ/m^2	890 x 10^3 Btu/ft^2
Suspension firing	550 GJ/m^3	14 x 10^3 Btu/ft^3
Fluidized bed	470 GJ/m^3	12 x 10^3 Btu/ft^3

[a]From Envirosphere (1980).

[b]1 h basis.

Rates of heat release are influenced both by fuel and by firing method. Governing heat release rates is a fundamental equation of Shafizadeh and DeGroot (1977):

$$I = dw(h)/dt \qquad\qquad (5\text{-}10)$$

where I is flame intensity, dw the change in weight, and dt change in time. Thus dw/dt is the rate of heat loss, and h the heat of combustion of the fuel. Dry wood burns at a higher flame temperature as has been shown in Fig. 5-3. Devolatilization occurs more rapidly as the dw/dt coefficient increases. Similarly, particle size influences the dw/dt term, with large particles having lower weight loss rates due to the insulative properties of wood.

More significantly, the dw/dt coefficients for various ranks of coal are lower than those associated with wood. This is due to a higher fixed-carbon content of coal associated with the more condensed structures. The high reactivity of wood fuel, relative to coal, as shown in the proximate analysis, is responsible for the relatively higher rate of heat release.

The practical results of the fuel influence are shown by a comparison of the wood and coal heat release rates when spreader-stokers are employed. The rate of heat release for wood is 10.4 GJ/m^2 of grate/h (890 x 10^3 Btu/ft^2/h). For coal, the rate is 8.8 GJ/m^2/h (750 x 10^3 Btu/h). This occurs despite the high moisture content of wood fuels (Envirosphere, 1980; Fryling, 1966). The influences of spreader-stoker and other firing methods on the rate of heat release with wood combustion are shown in Table IV. Values are for either grate surface or furnace volume, depending upon the firing method.

As can be seen from Table IV, wood combustion results in high rates of heat release. As a consequence, furnace volumes for wood-fired pile-burning (e.g., inclined-grate) systems and spreader-stokers are larger than comparable coal-fired systems. Such volumes are necessary to accommodate the flaming combustion that causes these rates of heat release.

IV.   CONCLUSIONS

The combustion of 100 J of wood fuel provides between 68 and 79 J of useful heat depending upon moisture content. Flame temperatures of 1400-1800 K (2100-2800°F) can be achieved. Heat release rates of 10.4 GJ/m^2 of grate/h (890 x 10^3 Btu/ft^2/h) can be obtained. These values demonstrate the utility of wood as a fuel in industrial settings. That utility is largely a result of the reactivity of wood fuel explained in Chapter 4.

It is significant to note that moisture content and excess air levels must be controlled carefully if wood fuel is to be a practical alternative to fossil fuels, however. The influence of both variables on combustion. If these variables are not carefully considered, combustion of wood is less than useful, as the quantity and quality of useful heat produced are severely impaired.

REFERENCES

Anderson, L. L. (1977). A wealth of waste, a shortage of energy, in "Fuels from Waste" (L. L. Anderson and D. A. Tillman, eds.), pp. 1-16. Academic Press, New York.

Arola, R. (1976). Wood fuels--How do they stack up, in "Energy and the Wood Products Industry." Proc. Forest Prod. Res. Soc., 34-41. Madison, Wisconsin.

Envirosphere (1980). "Program Negative Declaration for Biomass Demonstration Program of the California Energy Commission." Envirosphere Co., Newport Beach, California.

Fryling, G. R., ed. (1966). "Combustion Engineering." Combustion Engineering, Inc., New York.

Hill, P. G. (1977). "Power Generation: Resources, Hazards, Technology, and Costs." MIT Press, Cambridge, Massachusetts.

Hougen, O. F., Watson, K. A., and Razatz, R. A. (1954). "Chemical Process Principles, Part 1: Material and Energy Balances," 2nd ed. Wiley, New York.

Ince, P. J. (1977). "Estimating Effective Heating Value of Wood or Bark Fuels at Various Moisture Contents." USDA Forest Service General Tech. Rep. FPL, Madison, Wisconsin.

Junge, D. C. (1975). "Boilers Fired with Wood and Bark Residues." Res. Bull. 17, Forest Research Laboratory, Oregon State Univ., Corvallis.

Junge, D. C. (1979). "Design Guideline Handbook for Industrial Spreader Stoker Boilers Fired with Wood and Bark Residue Fuels." Oregon State Univ. Press, Corvallis.

Kester, R. A. (1980). "Nitrogen Oxide Emissions from a Pilot Plant Spreader Stoker Bark Fired Boiler." Ph.D. Thesis, University of Washington, Seattle.

MacCallum, C. (1979). The slooping grate as an alternative to the travelling grate on hog fuel fired boilers, in "Hardware for Energy Generation in the Forest Products Industry," pp. 53-62. Forest Products Research Society, Madison, Wisconsin.

Shafizadeh, F., and DeGroot, W. F. (1977). Thermal analysis of forest fuels, in "Fuels and Energy from Renewable Resources" (D. A. Tillman, K. V. Sarkanen, and L. L. Anderson, eds.), pp. 93–114. Academic Press, New York.

Tillman, D. A. (1978). "Wood as an Energy Resource." Academic Press, New York.

CHAPTER 6

AIRBORNE EMISSIONS FROM WOOD COMBUSTION

I. INTRODUCTION

In addition to useful heat, installations burning wood produce various quantities of airborne emissions such as particulates and oxides of nitrogen, liquid discharges such as boiler
blowdown, and solid wastes such as bottom ash and collected fly
ash. Liquid discharges from the fuel storage pile and the
boiler are readily controlled by wastewater treatment systems.
Similarly, solid wastes can be handled by properly designed
landfills or, in a few cases, by disposal on the land as a soils
amendment. Airborne emissions can also be controlled, but the
level of control makes these emissions the most critical environmental concern in wood combustion.

The basic types of airborne emissions of importance are particulates, carbon monoxide and hydrocarbons, and oxides of nitrogen. These are particularly important due to regulatory requirements imposed on wood combustors.

Because airborne emissons are the overriding environmental issue, they are examined here. This examination includes a discussion of the principles of emission formation, quantities of emissions expected, and methods for emission control.

Emphasis is placed on particulates, carbon monoxide and hydrocarbons, and oxides of nitrogen (NO_x). Sulfur dioxide is not considered to be a significant emission. Wood is generally a sulfur-free fuel, with typical concentrations being zero to less than 0.1 percent (see Chapter 2). Values for SO_2 reported in Tillman (1980), Kitto (1980), and Tillman and Jamison (1981) are negligible. Thus, this emission is not considered further. Beyond this point, particulates and NO_x are considered to be of most serious concern.

II. PARTICULATE FORMATION AND CONTROL

Particulates constitute the major source of emissions from hog-fuel-fired boilers. Generally, hog fuel consists mostly of bark and exhibits higher particulate emission rates than chipped or hammermilled wood when burned in a boiler. Sawdust has an intermediate value between bark and wood. This section discusses the amount, method of formation, and control measures necessary to reduce particulate emission from hog-fuel-fired boilers. In this discussion flyash is defined to include both flycarbon and inert materials, with flyash used interchangeably with particulate.

A. Quantities of Particulates Formed

Quantities of particulates formed are a function of inerts entering the furnace and, more importantly, firing practices. The differences between particulate emissions from a poorly run boiler and to a well run boiler can be quite substantial. Such a comparison is presented in Table I. For poor combustion, an excess air level of 130 percent was assumed and for good combustion 60 percent was employed. The units used are the conventional grams/standard dry cubic meter (g/SDCM)[grains/standard dry cubic foot (gr/SDCF)] with the introduction of units related to heat input, grams/megajoule (g/MJ), and lb/MBtu x 10^6. Since the feed rate into a large industrial boiler is generally not accurately known, alternative calculation methods have been developed to obtain mass/unit heat input quantities.

To obtain these values, an F factor is introduced, which has the units of SDCM/MJ or SDCF/Btu x 10^6 (U.S. Environmental Protection Agency, 1978). Calculation of the F factor is dependent on the ultimate analysis of the fuel and its higher heating value (HHV). The method of calculating the F factor for SI units is

$$F = \frac{10^3[(227.0\%)H + (95.7\%)C + (35.4\%)S + (8.6\%)N - (28.5\%)0]}{HHV \ J/kg}$$

$$= \frac{SDCM}{MJ} \tag{6-1}$$

and for English units is

$$F = \frac{10^6[(3.64\%)H + (0.53\%)C + (0.57\%)S + (0.04\%)N - (0.46\%)0]}{HHV \ (Btu/lb)}$$

$$= \frac{SDCF}{MBtu} \tag{6-2}$$

After calculating the F factor it can be applied for SI and for English units, respectively, to obtain

$$E = C_S F \frac{20.9}{20.9 - O_2(\text{percent})} \quad g/MJ \tag{6-3}$$

and

$$E = \frac{C_S F}{7006} \times \frac{20.0}{20.9 - O_2(\text{percent})} \quad \text{lb/Btu x } 10^6 \quad (6\text{-}4)$$

where C_S = g/SDCM and grains/SDCF, respectively.

As can be seen, the calculation of E depends on the percent excess air or percent oxygen (O_2). An approximate method for calculating the percent oxygen from the amount of excess air (EA), and for calculating excess air from the percent O_2 are given, respectively, by

$$\%O_2 = \frac{\%EA(0.264\%N_2)}{100 + \%EA} + \%CO_2 \quad (6\text{-}5)$$

$$EA = 100 \frac{\%O_2 - \%CO_2}{0.264\%N - (\%O_2 - CO/2)} \quad (6\text{-}6)$$

The percent oxygen from an Orsat analysis can also be used directly after being corrected to dry gas conditions. For wood-fired systems, the carbon monoxide (CO) term can be neglected.

Solving Eqs. (6-3) and (6-4) relates particulate emission rates to heat input. Using the F factors, it is possible to calculate the fuel feed rate (Q_H) in MJ/hr and Btu x 10^6 for a boiler by knowing the percent O_2 and the exhaust gas flow (Q_S) in SDCM/hr and SDCF/hr for metric and English units, respectively:

$$Q_H = Q_S(20.9 - \%O_2)/20.9F \quad \text{MJ/h} \quad (6\text{-}7)$$

$$= Q_S(20.9 - \%O_2)/20.9F \quad \text{MBtu x } 00^6/h \quad (6\text{-}8)$$

Thus, knowing Q_H, one can divide by the HHV (with respective units) and obtain the fuel feed rate to the boiler (kg/h or lb/h).

TABLE I. UNCONTROLLED EMISSION RATES FOR SPREADER-STOKERS[a,b]

| Fuel type | Poor combustion, maximum concentration | | Good combustion, minimum concentration | |
	$g/10^6$ J (lb/10^6 Btu)	g/SDCM (gr/SDCF)	$g/10^6$ J (lb/10^6 Btu)	g/SDCM (gr/SDCF)
Alder sawdust	3.52 (8.21)	6.04 (2.64)	1.21 (2.83)	3.02 (1.32)
Pine bark and douglas fir shavings	3.13 (7.15)	5.26 (2.30)	0.29 (0.68)	0.71 (0.31)
Alder bark	5.90 (13.7)	9.70 (4.24)	1.25 (2.89)	2.97 (1.30)
Ponderosa pine bark	21.21 (49.3)	34.90 (15.26)	1.77 (4.10)	4.21 (1.84)
Hemlock bark	5.34 (12.41)	8.78 (3.84)	0.91 (2.12)	2.17 (0.95)

[a] Adapted from Junge (1977).
[b] Poor and good combustion 130 and 60 percent excess air, respectively, using EPA-calculated F factors.

EPA has calculated F_d factors for wood- and bark-fired systems as 0.2478 SDCM/MJ (9280 SDCF/10^6 Btu) and 0.2588 SDCM/MJ (9640 SDCF/10^6 Btu), respectively, with an estimated error of 1.9 and 4.1 percent, respectively. The values in Table I are for a pilot plant scale spreader stoker located at Oregon State University. However, these values correspond reasonably well to values obtained from the U.S. Environmental Protection Agency (1979), and also values derived by Envirosphere Company (1980).

B. Mechanisms for Formation of Particulates

Particulate formation results from incomplete oxidation of the fuel, and from inerts in the fuel. Factors influencing their formation include fuel feed rate and characteristics of combustion air. After the noncarbon materials are stripped away from a molecule, carbon may not be oxidized to CO_2 due to temperature effects, retention time, and localized reducing conditions in the combustion chamber. These factors can be controlled by using proper fuel feed rates and proper levels of excess air and good air distribution.

1. Influences of Feed Rate

The fuel feed rate has a dominant influence on the amount of particulate carryover from the combustion zone to the particulate collection devices. This is shown in Fig. 6-1 for the spreader stoker facility at Oregon State University, using 50 percent excess air with 95 percent of that being over fire air (Junge, 1977). This phenomenon is a function of the inert ash entering the furnace and the degree to which carbon can be completely oxidized to CO_2. The time variable, essential to the completion of the complex char (solid-carbon particle) oxidation pathway, is minimized. Therefore, controlling the fuel rate to acceptable levels will enhance particulate control.

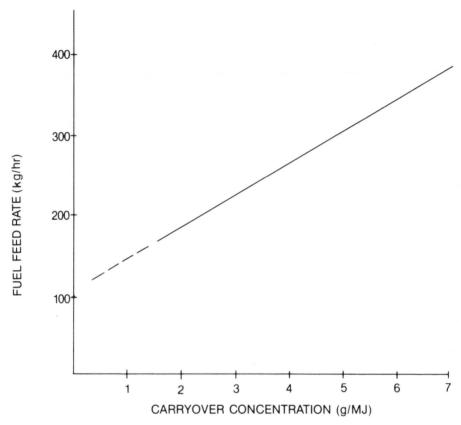

FIG. 6-1. Estimated concentration of suspended particulates
as a function of fuel feed rate (from Junge, 1977).

2.   Influence of Air Quantity and Distribution
     Figure 6-2 shows that increasing the percentage of excess
air will increase the particulate carryover when firing fines.
The reverse is true when firing larger size fuel particles below
75 percent excess air (Junge, 1977). However, when the excess
air levels increase above 75 percent, carryover will then increase due to higher furnace velocities. Consequently, adjust-
ing the level of excess air in response to the particle size,
firing method, and fuel moisture content will help reduce particulate emissions.

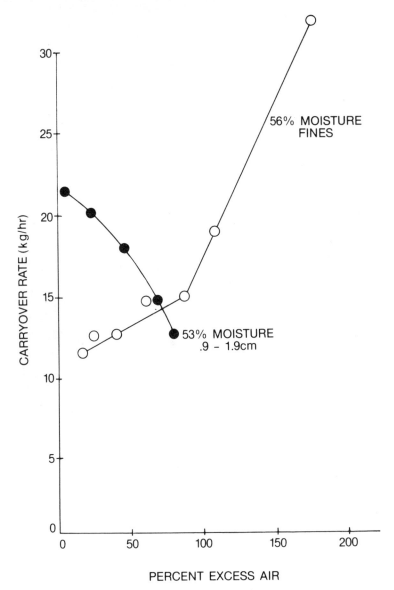

FIG. 6-2. Comparison of carryover rate versus excess air for Douglas fir bark and wet bark fines (from Junge, 1977).

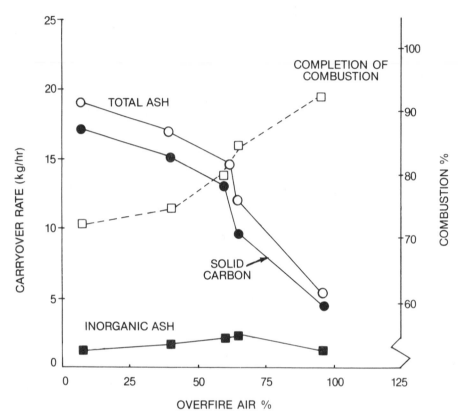

FIG. 6-3. Carryover rate and percent combustion as a function of overfire air percentage (from Junge, 1977).

Figure 6-3 shows the relationship between carryover, overfire air, and percent completion of solid carbon combustion,
using 50 percent excess air for a 50 percent MC_G fuel, at the
Oregon State facility. It can be seen that as the amount of
overfire air increases above 75 percent, solid carbon conversion
approaches 95 percent, with the total carryover being substantially reduced. With these conditions the particulate formed
has a lower ratio of carbon to ash than when using more underfire air. With the overfire air levels raised to 95 percent
(underfire air 5 percent, excess air 50 percent) the combustion
takes place in two distinct flame fronts, with limited amounts

of char burning in suspension (Junge, 1978). The char tends to remain on the grate, completing glowing oxidation, while the volatiles are burned in the vicinity of the overfire air ports. This staged approach to combustion may be applied to other firing methods such as pile burning, although it is more difficult to achieve with suspension firing.

3. Characteristics of the Wood Particulate Formed

Particulates generally range in size from less than 10^3 mm (submicron) up to 2 mm (Junge, 1979). The size of the particulate is dependent on the combustion conditions used and the size characteristics of the hogged fuel. Table II presents particulate size data from the pilot plant scale spreader stoker located at Oregon State University. Note the large size of the particulate and the amount of variation. Table III presents density characteristics of particulate, which range from 0.07 g/cm^3 (4.4 lb/ft^3) to 0.36 g/cm^3 (22.5 lb/ft^3), also presented is the percent combustibles in the particulate.

Generally, as the percent combustibles decrease, the density increases due to the increasing inerts content. Because the carbon content of the total particulate emissions may be high, it is periodically reinjected back into the system to recover some of its heat value. Reinjection of this material increases the mass mean concentration of small-particulate particles, thus decreasing the efficiency of the collection equipment. This recirculation also tends to erode boiler tubes and heat transfer surfaces due to the abrasive nature of the inorganic ash content of the reinjected flyash.

The values given in Table II are a geometric mass (weight) average representation of particulate size distribution. Using a number average distribution, the particulate less than 10^3 mm would play a major role. This small material increases opacity and presents collection problems for conventional multiclone systems, thus creating needs for more efficient collection devices.

TABLE II.   GEOMETRIC MASS MEAN PARTICLE SIZE (mm)[a,b]

Species	Average	Minimum	Maximum
Douglas fir bark	0.68	0.06	1.80
Douglas fir shavings	0.74	0.51	0.86
Alder sawdust	0.76	0.46	1.05
Alder bark	0.15	0.05	0.48
Ponderosa pine bark	1.40	0.89	1.95
Hemlock bark	0.78	0.24	1.25

[a]From Junge (1979).
[b]1 mm = 1000 microns.

TABLE III.   GEOMETRIC MASS MEAN PARTICLE SIZE (mm)[a]

Species	Density of particulate g/cm^3 (lb/ft^3)		Combustible content percent
Douglas fir bark	0.137	8.55	NA
Douglas fir shavings	0.097	6.05	91.26
Alder sawdust	0.083	5.18	72.50
Alder bark	0.358	22.34	20.29
Ponderosa pine bark	0.077	4.80	89.70
Hemlock bark	0.109	6.80	70.05

[a]From Junge (1979).

C. Methods for Control

Control methods are divided into two categories: (1) internal methods, consisting of improved combustion through better control of furnace parameters and ultimately the combustion chemistry; and (2) postcombustion control devices, including cyclones, baghouses, scrubbers, and electrostatic precipitators.

1. Internal Methods

Control of the fuel feed rate is of initial importance. Most older systems feed the fuel intermediately when the fuel pile reaches a certain level, thus causing fuel-rich conditions due to the sudden increase in pile volume. A better method is to continuously feed the fuel to the furnace using screw conveyors from surge bins. This serves three purposes. It reduces the frequency of fuel rich conditions, it makes the system more responsive, and it reduces the uncontrolled flow of excess air through the fuel feed system, since a solid plug of fuel is continuously fed. It is also important to investigate the steam profile of the facility and even out the steam load so there are no sudden surges imposed on the furnace. This can be accomplished by using electronic steam-integrating chart recorders. The electronic signal from the steam integrater and pile height indicators can be used to control the fuel feed rate.

Control of excess air levels and air distribution is also of importance as can be seen from Fig. 6-2. Before proper levels of excess air can be maintained, sources of unwanted air should be eliminated. Once this is accomplished, electronic oxygen meters can be installed to control the damper settings of the induced or forced draft fans. The next step is to control proper levels of overfire and underfire air to achieve staged combustion. This is most easily accomplished in pile-burning and spreader-stoker systems. This will allow for good separation of glowing and flaming combustion as discussed in Chapter 4, and reduce particulate carryover. It is also important to cover the grates completely with fuel so that all the underfire air goes through the pile, not around it. These changes or controls can be readily designed into new systems. In making these changes or in designing new systems, the above discussion should be related to staged combustion.

2. Postcombustion Devices

Postcombustion emission control devices are needed to reduce and maintain boiler emissions below 0.23 g/m^3 (0.1 gr/SDCF). Generally, multiple cyclones (multiclones) remove particulates of greater than 0.005 mm (5 microns) with an efficiency of 80-90 percent, but for particulates less than 0.002 mm (2 microns) efficiencies become unacceptable. Other options available include wet scrubbers, baghouses, dry scrubbers, and electrostatic precipitators.

Reduction of uncontrolled particulate matter to 0.45 g/SDCM (0.2 grains/ SDCF) or below can generally be achieved through the use of a multiclone collector and good furnace operation. However, further reductions to .23 g/SDCM (.1 grains/SDCF) can be accomplished by modifications in the combustion chemistry, through better control of furnace parameters, and through such collection devices as scrubbers, baghouses, and electrostatic precipitators. These are discussed below.

Wet scrubbers can consist of spray towers of trays and packed bed systems. They are generally not recommended unless SO_2 removal is a significant problem. Baghouses are fabric filters. Baghouses have very high collection efficiencies and provide for the dry handling of the collected particulate. Today bag technology limits the input temperature to approximately 490 K (425°F). As a result, some type of gas cooling is needed, such as economizers or water sprays. However, baghouse temperatures must be maintained high enough to prevent condensation and corrosion problems.

Dry scrubbers use a moving granular filtration media such as pea-gravel to collect particulates. The moving bed also provides a self-cleaning action of the filtration media. Collection efficiency is high and can be influenced by control of the recirculation rate of the media and by an electrostatic charge in the scrubber.

Electrostatic precipitator efficiency is a function of the corona power. However, the resistivity of flyash impedes effective precipitator operation, and usually some modification of the resistivity is needed.

In most cases, multiclones should be used before the collection devices described above to remove larger particulate and sparks, thus improving secondary collection efficiencies. Typical efficiencies pressure drops, and particle size data for various particulate collection devices, properly designed and operated, are presented in Table IV.

TABLE IV.   CHARACTERISTICS OF COLLECTION SYSTEMS[a]

	Pressure drop (in. w.c.)	Collection efficiency (percent weight)	Particle size range micrometers (microns)
Electrostatic precipitation[b]	0.2-1	80-99	>2
Baghouse	4-6	95-99	>0.2
Dry scrubber	6-12	90-99	N/A
Wet scrubber[c]	6-10	60-90	>5
Multiclones	2-10	80-90	>5

[a]Adapted from Cavaseno et al. (1980) and NATO/CCMS (1973).
[b]Includes irrigated and nonirrigated.
[c]Does not include venturi.

III. CARBON MONOXIDE AND HYDROCARBON EMISSIONS

The emission of CO and HC is largely the result of incomplete combustion. This can be directly attributed to a low carbon conversion efficiency to carbon dioxide (CO_2) and improper air turbulence and distribution. Carbon monoxide can also be formed by the high-temperature dissociaton of CO_2. Thus, methods of CO and HO formation and control are discussed below.

A. Typical Quantities of CO and HC Formed by Wood Combustion

Measured carbon monoxide emission rates have ranged from 0.3 to 35 g of CO/kg of fuel burned (U.S. Environmental Protection Agency, 1979), although lower levels are readily achievable. For a typical wood fuel (oven dry basis), the above values correspond to 0.0146 g/MJ (0.034 lb/10^6 Btu) and 1.7 g/MJ (4.0 lb/10^6 Btu). The higher values reflect poor combustion practice. There seems to be no correlation between CO emission rates and boiler size, or with co-firing of wood and fossil fuels.

The formation of CO and HC can result as a consequence of incomplete combustion. CO is formed by the partial oxidation of carbon, while HC results from unoxidized chemical fragments. This is a consequence of inadequate retention time in the combustion zone, which can be caused by high levels of excess air increasing the velocity through the fuel pile and combustion zones. Also, low levels of excess air can cause fuel-rich conditions and incomplete oxidation of the fuel, with pile thickness also being related to incomplete oxidation. Improper air distribution can produce reducing conditions, generally occurring in pockets in the combustion zone.

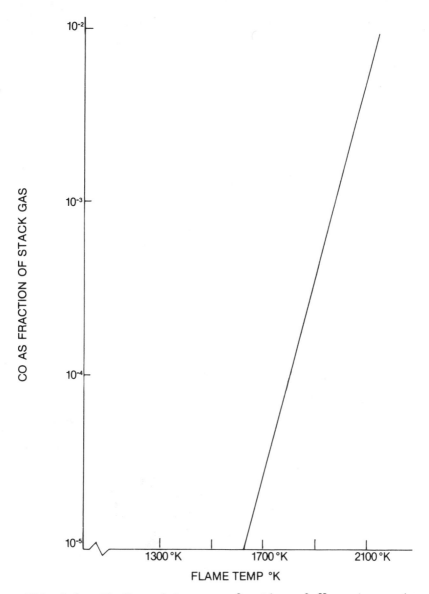

FIG. 6-4. CO dissociaton as a function of flame temperature at equilibrium conditions. The basis for these calculations assumes complete oxidation of carbon to CO_2 and then dissociation of CO_2 to CO. It also assumes achieving equilibrium. Since equilibrium is rarely achieved, these may be treated as maximum values for CO generated by dissociation.

Carbon monoxide can also be formed by the disassociation (reduction) of carbon dioxide at high temperatures:

$$CO_2 + \Delta H \longrightarrow CO \qquad\qquad (6-9)$$

For this to occur at all the flame temperature must exceed 1500 K (2240°F). Figure 6-4 shows CO formation as a function of flame temperature at equilibrium. An increase in temperature from 1600 to 1800 K (2420 to 2800°F) produces approximately an order of magnitude change in CO concentration from CO_2 dissociation (L. L. Anderson, private communication 1980). Wood fuel combustion typically occurs in the range 1500-1800 K (2200-2800°F) for grate-fired systems, which corresponds to a concentration from dissociaton of CO of less than 100 ppm, even when equilibrium is achieved.

B. Methods of Control

Important parameters in controlling CO and HC emission are to maintain proper excess air levels and obtain a high carbon conversion efficiency to carbon dioxide, typically in excess of 98 percent for a well-run boiler. It is also important to prevent boiler surging caused by sudden increase or decreases in steam demand, particuarly when firing wet hog fuel. Excess air and CO/HC can be measured by electronic instrumentation. By measuring both parameters, excess air can be adjusted while minimizing CO/HC concentrations. Measurement and control to <10 ppm has been achieved.

IV. EMISSIONS OF OXIDES OF NITROGEN FROM WOOD COMBUSTION

Numerous oxides of nitrogen can be considered, including those identified in Table V. Of the seven oxides of nitrogen listed there, NO and NO_2 are the most prevalent. Over 90 per-

TABLE V.   MAJOR OXIDES OF NITROGEN[a]

	Chemical formula	Molecular weight
Nitrogen oxide	NO	30
Nitrogen dioxide	NO_2	46
Nitrogen tetroxide	N_2O_4	92
Nitrous oxide	N_2O	44
Nitrogen sesquioxide	N_2O_3	76
Nitrogen pentoxide	N_2O_5	108
Nitrogen trioxide	NO_3	62

[a]From Kester (1980).

cent of the NO_x produced by wood combustion is NO, with the rest being most exclusively NO_2 (Kester, 1980).

Nitrogen oxides are considered to be important pollutants because they are precursors to ozone (O_3), and thus to photo-chemical smog. Further, they have been shown to be contributors to acid rain. Because they are considered important, emission factors have been developed to estimate the level of NO_x emissions from wood-fired facilities. The USEPA publication AP-42, for example, shows an emission factor for wood-fired spreader-stokers of 4.1 kg/tonne of fuel (10 lb/Ton) as NO_2. Kester (1980) has shown this factor to be high by almost an order of magnitude. The Kester factor is 0.45 kg NO_x/Tonne of fuel (1.1 lb/Ton), based on 94 specific tests on the Oregon State University experimental spreader-stoker. Kester's value converts to 0.045 kg/GJ (0.11 lb/10^6 Btu).

While the Kester factor approximates NO formation from the combustion of Douglas fir bark in a spreader-stoker, research in fossil fuels show that there are numerous interdependent variables to be considered in NO_x formation, including (1) flame temperature, (2) level of excess air, (3) fuel nitrogen content, and (4) firing method. Two types of NO_x must be considered: thermal NO_x and fuel NO_x.

A. Formation of Thermal NO_x

Fixation of nitrogen that enters the combustor as part of the combustion air causes thermal NO_x. The process is not fuel specific, for thermal NO_x is produced in the combustion of all fuels from natural gas to coal. The formation of thermal NO_x is largely driven by flame temperature (thus its name). Associated variables include the level of excess air, particularly as it influences flame temperature and the relative concentration of nitrogen in the reactants.

Typically, the formation of thermal NO_x is shown by the reaction

$$N_2 + O_2 \longrightarrow 2NO \tag{6-10}$$

While Eq. (6-10) is a useful summary statement, it represents a four-centered reaction that probably does not occur (Palmer, 1974). Rather, the classific Zeldovich mechanism is usually used to explain NO formation (Palmer, 1974):

$$O + N_2 \longrightarrow NO + N \tag{6-11}$$

$$N + O_2 \longrightarrow NO + O \tag{6-12}$$

In addition to formation of NO_x from the atom shuttling Zeldovich mechanism, some NO_x is formed as follows:

$$N_2 + O_2 \longrightarrow N_2O + O \tag{6-13}$$

$$N_2O + O \longrightarrow 2NO \tag{6-14}$$

$$N + OH \longrightarrow NO + H \tag{6-15}$$

$$N_2 + OH \longrightarrow N_2O + H \tag{6-16}$$

Reaction (6-16) would then be followed by reaction (6-14) (Palmer, 1974). In addition to these reactions, Fenimore (1971) has identified the "prompt NO" formation mechanism for hydrocarbon fuels:

$$N_2 + CH \longrightarrow HCN + N \qquad\qquad (6-18)$$

Reaction (6-17) would then be followed by reaction (6-12).

Once formed, the NO can be consumed by the reverse of reactions (6-11) and (6-12), which are also highly temperature dependent. Thus, rapid cooling will tend to "freeze" the NO_x concentration in the final stack gas (Palmer, 1974).

The influence of flame temperature and excess air levels on thermal NO_x formation can be seen from calculations of the equilibrium concentrations of NO in stack gases of wood combustion. Figure 6-5 shows the influence of flame temperature and Fig. 6-6 shows the influence of excess air. Calculations were made by the NASA combustion model introduced in Chapter 6, assuming nitrogen-free wood fuel.

As a practical matter, equilibrium is rarely, if ever, reached. Concentrations may be as low as an order of magnitude below equilibrium concentrations as the Kester (1980) data show. However, in addition to showing flame temperature dependence, these data show that maximum NO concentrations are reached at ~10 percent excess air levels. There is a confluence of temperature and nitrogen concentration factors at that point. These data lead to the conclusion that staged combustion, with substoichiometric quantities of primary air and substantial levels of excess secondary air, can be used effectively to control formation of thermal NO in wood combustion. Some flame temperature is sacrificed in the process, however.

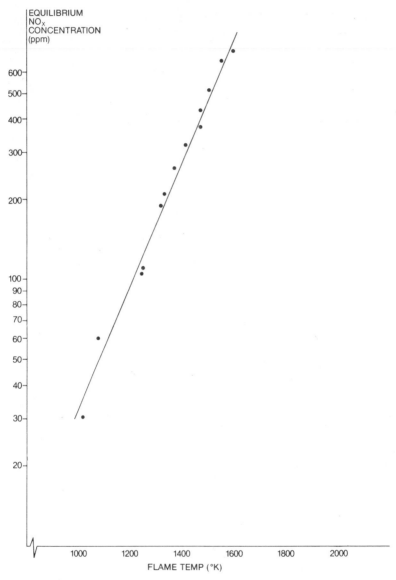

FIG. 6-5. NO concentrations in wood stack gas as a function of flame temperature, assuming complete oxidation of carbon to CO_2 and assuming achieving equilibium. Note again that equilibrium is rarely achieved. Thus, these values may be considered as maxima for thermal NO_x produced by wood combustion.

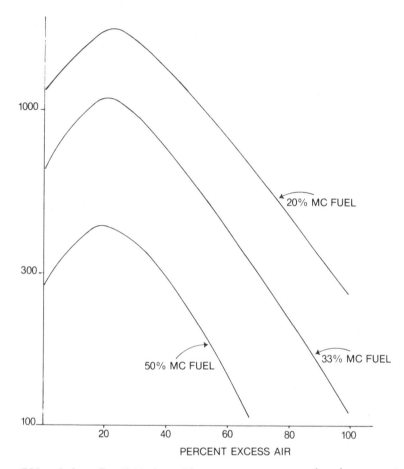

FIG. 6-6. Equilibrium NO_x concentrations (ppm) as a function of percent excess air for wood fuel combustion. Note that maximum values are achieved at ~20 percent excess air. As fuel air mixtures become increasingly rich or increasingly lean from that point, NO_x concentrations decline. These values assume 100 percent carbon conversion to CO_2 and the achieving of equilibrium in the combustor. Again, these may be treated as threatened values unlikely to be achieved or even approached in practice (from L. L. Anderson, private communication, 1980).

B. Formation of Fuel NO_x

Wood fuels typically are low in nitrogen content, with values ranging from 0.07 to 0.2 percent. However, wood fuels such as red alder bark may contain upwards of 0.4 percent N (Smith, 1981). Thus, formation of fuel NO_x may be significant, particularly in light of certain proposals to grow red alder as a fuel material (see Inman et al., 1977; Henry, 1979). Other biomass fuels such as rice hulls and straws also contain relatively high (>0.5 percent) nitrogen concentrations as well.

Understanding the formation of NO from fuel nitrogen involves an extension of the combustion model presented in Chapter 4. While the specifics remain undefined, sufficient data have been developed with respect to fossil fuels that a general understanding can be achieved.

The major influences on fuel NO_x formation have been identified as follows: (1) nitrogen content of the fuel (Pershing et al., 1978); (2) firing method (Giammar et al., 1980; D. W. Pershing, personal communication, October 29, 1980); and (3) firing conditions (Beer et al., 1980a, b). Flame temperature may have a minor influence on NO_x formation under specific conditions (Adams et al., 1980; Vogt and Laurendeau, 1978; Pershing et al., 1978). Applying these influences to biomass requires use of both oil and coal studies.

1. Fuel Nitrogen Content Influences

The quantity and form of fuel nitrogen both influence NO_x formation. The quantity of fuel nitrogen has been identified previously for wood, the nitrogen is largely contained in amine functional groups, which are contained in the protein content of the inner bark (Cowling and Kirk, 1976). Trace quantities of nitrogen also are contained in lignin precursers (Sarkanen, 1971). In contrast, nitrogen in coal is contained in amines, pyridines, carboyoles, quinolines, and pyrroles (Vogt and

Laurendeau, 1978). Thus, the nitrogen in the wood fuel is in a more accessible, reactive form.

The amine functional groups are volatilized readily in the solid-phase pyrolysis zone. Amine radicals may be produced by homolytic cleavage of C-N bonds. These radicals may be stabilized in precombustion gas phase reactions, if sufficient retention time exists in such zones. Ultimately, they may be reduced to N_2 under such conditions. If passed into the primary combustion zone in highly reactive form, they may react with hydroxy radicals. In such cases one can speculate that hydrogen atoms from the amine radical may be abstracted as:

$$NH_2 + OH \longrightarrow NH + H_2O \qquad\qquad (6\text{-}18)$$

Fuel NO_x studies typically show percentages of converting fuel N to NO_x. These conversions are based on moles of fuel N to moles of NO or NO_2. For oil firing, the range of conversion percentages is about 55-80 percent (Pershing et al., 1978). Coe (1978) has documented this relationship, with data fitting the following linear expression:

$$NO_x \text{ (ppm)} = 275 \text{ (percent } N_{fuel}) \qquad\qquad (6\text{-}19)$$

Coe employs 15 percent excess air in the system. At 0.5 and 1 percent nitrogen in the fuel, NO_x concentrations are 138 and 275 ppm, reespectively.

The use of oil based data is appropriate since nitrogen in oil typically is in volatile form. However, oil firing is analogous to suspension firing. Because wood is fired in pile burning and spreader-stoker systems, additioi.al data are necessary.

2. Firing Method Influences

Data on fuel nitrogen conversion from a coal combustion as a function of firing method have been developed by Giammar et al. (1980), Pershing and Wendt (1976), and D. W. Pershing (personal

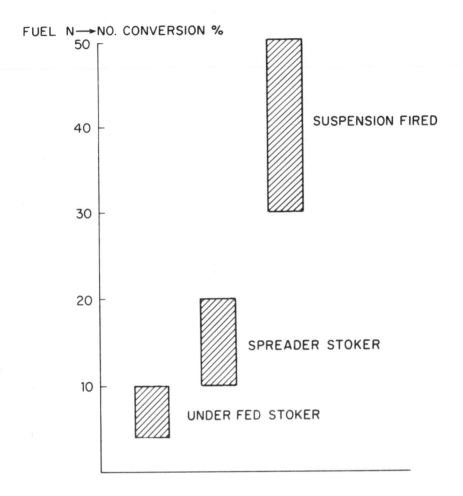

FIG. 6-7. Fuel N conversion as influenced by firing method, measured on a percentage basis. These values may be conservative for pile burning of wood fuel as the reactive fuel-bound nitrogen and wood may be volatilized, and then reduced in precombustion reactions to N_2. Alternatively, they may be optimistic for suspension burning, again due to the reactivity of the nitrogen bound in the wood fuel. These data are presented for coal combustion (from Giammar et al., 1980; D. W. Pershing, personal communication, October 29, 1980).

communication, October 19, 1980). These are shown in Fig. 6-7. For underfed stokers, conversion percentages are low, typically <10 percent; and for spreader-stokers, where some combustion occurs in suspension, conversion is 10-20 percent (Giammar et al., 1980). For suspension firing, fuel nitrogen to NO_x conversion percentages in excess of 50 percent have been measured (Pershing and Wendt, 1976; D. W. Pershing, personal communication, January 29, 1980). To understand these values, it is essential to examine the related parameter, excess air conditions.

The previous section shows that increasing the availabiity of oxygen to the fuel in a reactive state causes the conversion of fuel nitrogen to NO_x. As more oxygen is available in the initial gas phase reaction zone, more radicals undergo immediate oxidation reactions rather than reduction (e.g., to N_2) reactions. If oxygen is highly available, oxidation reactions will tend to dominate.

In underfed stokers, where no suspension firing occurs, the bed can be maintained at substoichiometric conditions. The supply of oxygen can be very limited, and reducing reactions can dominate. In spreader-stoker firing, it is more difficult to maintain substoichiometric conditions at all times around incoming fuel particles. Some oxygen is immediately available to the initial volatiles produced in the combustor. In suspension burning, an ample supply of oxygen is usually available immediately as the fuel is carried into the combustor in an airstream. Strict staging of combustion reactions is more difficult to accomplish; and complete separation of flaming (gaseous) and glowing (char) combustion is virtually impossible. Thus, the interplay of firing methods and firing conditions exerts a dominant influence on the conversion of fuel N to fuel NO_x.

Because wood fuel is more reactive than coal, it is believed that these conversion percentages may be somewhat low when representing total NO_x emission. Depending on air distribution, one might expect wood fired spreader-stokers to show conversion percentages in the 15-25 percent range. For example, if one assumes that the Kester (1980) factor is attributable entirely to fuel N, then the conversion percentage is 18 percent.

The Kester data show NO_x concentrations from 25 to 103 ppm. Further, the analysis shows NO_x emissions to be influenced by fuel feed rate, flame temperature, level of excess air, and--to a lesser extent--fuel moisture content and particle size. The coefficient of determination (r^2) for fuel feed rate is 0.42 and the coefficient of determination for flame temperature is 0.30. Thus, the 1.1 lb NO_x/ton burned contains some thermal NO_x as well as fuel NO_x. Given those values one can assume a fuel N $\longrightarrow$ NO conversion percentage in the 10-15 percent range for the Kester experiments.

One would expect a relatively low conversion due to the controls available on the spreader-stoker employed and the emphasis on staged combustion (hence, the promotion of heterogeneous reduction reactions) in the experiments [see Junge (1977) for description of program].

3. Influence of Flame Temperature

Flame temperature is far less important in converting fuel N to NO_x than producing thermal NO_x. Glassman (1977), for example, makes the following statement:

> When fuel nitrogen compounds are present, high NO yields are obtained for lean and stoichiometric mixtures and relatively lower yields are found for fuel-rich mixtures. The NO yields appear to be only slightly dependent on temperature and thus indicate a low activation energy step.

Haynes and Kirov (1974) found high (370 ppm) NO concentrations in low-temperature coal combustion while gas phase oxidation was occurring, a phenomenon caused by fuel-bound nitrogen in the coal. Alternatively, Adams et al. (1980) report that low-temperature flameless catalytic combustion is ineffective in reducing NO formation from fuel-bound nitrogen.

There is some evidence that flame temperature is influential, however, under certain circumstances. At very high temperatures where $T \geq 2590$ K ($4200°F$), Wendt and Pershing (1977) show fuel NO_x conversion increases. There may be a very low temperature phenomenon also, although opinion on this is divided. Vogt and Laurendeau (1978) show stable conversion percentages of fuel bound nitrogen from coal at 20 percent below 1000 K ($1400°F$). They show a linear increase from 20-70 percent as temperature increases to about 1310 K ($\sim1900°F$). Significant to note, however, is the fact that the low temperature volatiles are amines (Vogt and Laurendeau, 1978). Temperatures reported are furnace wall rather than flame temperatures.

There is other support in the literature for this lower temperature effect. Pereira et al. (1974) show increasing NO formation in the temperature region 866-1033 K ($1100-1400°F$), and a leveling off at that point. Temperature reported here is bed temperature in a fluidized-bed combustor. Again, the form of the nitrogen in the coal is given as a reason for this phenomenon.

DeSoete (1974) provides perhaps the clearest explanation of firing method and flame temperature influences on the conversion of fuel N to NO_x. He states that in the experiments on liquids and gaseous fuels which he performed, nitrogen was added to the fuel in the form of amines, ammonia, cyanogen, and NO itself. The final NO_x concentrations were particularly influenced by oxygen concentrations, hence the availability of oxygen to volatile nitrogen species. Temperature played a very small role in NO_x formation from the fuel bound nitrogen, and the temperature primarily affected the formation of radical nitrogen species. Temperature influence was also affected by equivalence ratio or level of excess air.

DeSoete's experiments are largely analogous to suspension firing of biomass fuels where no staged combustion typically occurs, and where nitrogen from the fuel is readily accessible to copious quantities of oxygen in the combustion air stream. They are less applicalbe to grate fired systems where staged combustion can be readily achieved in significant measure.

4. Fuel NO_x Control

Control of fuel NO_x formation largely involves the staging of combustion to create a significant opportunity for reduction reactions in the precombustion gas phase reaction zone. This is easily accomplished by the placement of overfire air ducts and use of overfire air to carry all excess air in pile-burning and spreader-stoker systems. It is accompished by maintaining fuel-rich conditions on the bed of the combustor, and achieving a large physical separation between the char oxidation zone and the flaming combustion zone.

In staged combustion with pile burning and spreader-stoker equipment, it may be advantageous to have the nitrogen in reactive form. If the nitrogen is released in volatile form to a substantial precombustion gas phase reaction zone, gas phase reduction reactions may predominate. The fuel N ultimately would come off largely as N_2. The fuel N remaining in the char ultimately would be oxidized by undergrate air in glowing combustion. Higher conversion percentages could result.

In suspension burning, complete staging of combustion is more difficult to achieve. Because the fuel is suspended in a stream of combustion air, oxygen is typically immediately available to the volatiles. Hence, higher fuel N $\longrightarrow$ NO conversion percentages are virtually inevitable. In this case, volatile nitrogen would be a detriment because O_2 would be immedately available to amine radicals. Reducing reactions would have less of a change to occur.

Other control strategies are also promoted for control of both thermal NO_x and fuel NO_x. These include flue gas recirculation and ammonia injection. Such strategies work to the extent that they promote reduction reactions. The practicality of such systems is very site specific.

Staged combustion with some control of flame temperature appears to give the highest potential for NO_x control. Given the relatively low fuel N content for most species (see Chapter 3), NO_x emissions do not appear particularly problematical. Difficulties could arise, however, if nitrogen fixing species were grown to be used for fuel in a relatively large (e.g., 50 MW) powerplants.

V. CONCLUSIONS

The chemical properties of wood, which govern the pathways of its combustion, result in emissions of particulates, CO/HC, SO_x, and NO_x. These emissions are quite modest when compared to fossil fuels. Particulate emissions can be controlled to 0.009 g/MJ (approximately 0.02 lb/10^6 Btu), CO/HC emissions can be maintained well below such levels, and NO_x (as NO_2) emissions can be held to levels of approximately 0.05 g/MJ (0.12 lb/10^6 Btu).

Variables governing the formation and control of such effluents largely involve use of advanced furnace design and proper operation. Additionally, stack gas clean-up equipment is employed to achieve appropriate levels of emissions, particularly flyash.

Because wood fuels are relatively benign with respect to fossil fuels and other biomass forms, they are often considered in proposals for electric power generation within and outside the forest products industry. Thus, such proposals are now considered.

REFERENCES

Adams, B., et al. (1980). "Flameless Fire: The Practical Potential of Catalytic Combustion." Draft Report, National Academy of Sciences, Washington, D.C.

Beer, J. M., et al. (1980a). Two phase processes in the control of nitrogen oxide formation in fossil fuel flames, Proc. Joint Symp. Stationary Combustion NO_x Control, Vol. 4. EPA/EPRI.

Beer, J. M., et al. (1980b). Control of NO_x and particulates emission from SRC-II spray flames, Proc. Joint Symp. Stationary Combustion NO_x Control, Vol. 4. EPA/EPRI.

Casaveno V., et al., eds. (1980). "Industrial Air Pollution Engineering." McGraw-Hill, New York.

Coe, W. W. (1978). Combustion: Efficiency vs. NO_x, Hydrocarbon Processing, 130-134.

Cowling, E. B., and Kirk, T. K. (1976). Properties of cellulose and lignocellulosic materials as substrates for enzymatic conversion processes, in "Enzymatic Conversion of Cellulosic Materials: Technology and Applications" (E. G. Gaden, M. H. Maudels, E. T. Reese, and L. A. Spano, eds.), pp. 95-124. Wiley (Interscience), New York.

DeSoete, G. G. (1974). Overall reaction rates of NO and N_2 formation from fuel nitrogen, Proc. 15th Int. Symp. Combustion. The Combustion Institute, Pittsburgh, Pennsylvania.

Envirosphere (1980). "Program Negative Declaration for Biomass Demonstration Program of the California Energy Commission." Envirosphere Co., Newport Beach, California.

Fenimore, C. P. (1971). Formation of nitric oxide in pre-mixed hydrocarbon flames, Proc. 13th Int. Symp. Combustion. The Combustion Institute, Pittsburgh, Pennsylvania.

Giammar, R. D., et al. (1980). Evaluation of emissions and control for industrial stoker boilers, Proc. Joint Symp. Stationary Combustion NO_x Control, Vol. 3. EPA/EPRI.

Glassman, I. (1977). "Combustion." Academic Press, New York.

Haynes, B., and Kirov, N. Y. (1974). Nitric oxide formation during the combustion of coal, Combustion Flame 23:277-278.

Henry, J. F. (1979). The silvicultural energy farm in per-spective, in "Progress in Biomass Conversion," Vol. 1 (K. V. Sarkanen and D. A. Tillman, eds.), pp. 215-256. Academic Press, New York.

Inman, R. E., et al. (1977). "Silvicultural Biomass Farms, Vol. 4: Site-Specific Production Studies and Cost Analyses." MITRE Tech. Rep. No. 7347, U.S. Dept. of Commerce, Washington, D.C.

Junge, D. C. (1975). "Boilers Fired with Wood and Bark Re-sidues." Res. Bull. 17, Forest Research Laboratory, Oregon State Univ., Corvallis.

Junge, D. C. (1977). "Investigation of the Rate of Combus-tion of Wood Residue Fuel." Tech. Progress Rep. No. 1. Forest Research Laboratory, Oregon State University, Corvallis.

Junge, D. C. (1978). Wood combustion--Results and applica-tions of recent research, Proc. Conf. Feb. 20-21, 1978.

Junge, D. C. (1979). "Design Guideline Handbook for Indus-trial Spreader Stoker Boilers Fired with Wood and Bark Residue Fuels." Oregon State Univ. Press, Corvallis.

Kester, R. A. (1980). "Nitrogen Oxide Emissions from a Pilot Plant Spreader Stoker Bark Fired Boiler." Ph.D. Thesis, University of Washington, Seattle.

Kitto, W. D. (1980). Environmental considerations in wood fuel utilization, in "Progress in Biomass Conversion," Vol. 2

(K. V. Sarkanen and D. A. Tillman, eds.). Academic Press, New York.

NATO/CCMS (1973). "Control Techniques for Particulate Air Pollutants." No. 13 prepared by Expert Panel for Air Pollution Control Technology, Committee on the Challenges of Modern Society, NATO.

Palmer, H. B. (1974). Equilibria and chemical kinetics in flames, in "Combustion Technology: Some Modern Developments" (H. B. Palmer and J. M. Beer, eds.), pp. 2-33. Academic Press, New York.

Pereira, F. J., Beer, J. M., Gibbs, B., and Hedley, A. B. (1974). NO_x emissions from fluidized-bed coal combustion, Proc. 15th Int. Symp. Combustion.

Pershing, D. W., and Wendt, J. (1976). Pulverized coal combustion: The influence of flame temperature and coal composition on thermal and fuel NO_x, Proc. 16th Int. Symp. Combustion.

Pershing, D. W., et al. (1978). The influence of fuel composition and flame temperature on the formation of thermal and fuel NO_x in residual oil flames, Proc. 17th Int. Symp. Combustion.

Sarkanen, K. V. (1971). Precursors and their polymerization, in Lignins: Occurrence, Formation, Structure, and Reactions" (K. V. Sarkanen and C. H. Ludwig, eds.), pp. 95-164. Wiley (Interscience), New York.

Smith, W. R. (1981). Unpublished research conducted at the Wood Fuels Laboratory, Department of Forest Resources, Univ. of Washington, Seattle.

Tillman, D. A. (1980). Fuels from waste, in "Kirk-Othmer Encyclopedia of Chemical Technology," Vol. 2, 3rd ed. Wiley, New York.

Tillman, D. A., and Jamison, R. L. (1981). A review of cogeneration with wood fuel, Fuel Proc. Technol., in press.

.

U.S. Environmental Protection Agency (1978). APTI Course 450: Source Sampling for Particulate Pollutants.

U.S. Environmental Protection Agency (1979). Unpublished draft report, received November 5 for review.

Vogt, R. A., and Laurendeau, N. M. (1978). Effect of devolatilization on nitric oxide formation from coal nitrogen, Fuel 57:232-234.

Wendt, J., and Pershing, D. W. (1977). PhysicaL mechanisms governing the oxidation of volatile fuel nitrogen in pulverized coal flames, Combustion Sci. Technol. 16:111-121.

Wendt, J., Pershing, D. W., and Glass, J. W. (1978). Pulverized coal combustion: NO_x formation mechanisms under fuel rich and staged combustion conditions, Proc. 17th Int. Symp. Combustion.

CHAPTER 7

ELECTRICITY GENERATION:

THE EFFICIENCY OF WOOD FUEL UTILIZATION

I. INTRODUCTION

Currently wood fuels are used to raise process steam for lumber and pulp production, and direct process heat for use in plywood manufacture. Because such use is compatible with electricity production, a considerable quantity of wood is used to support the cogeneration of electricity and process steam. Conversely, a minor fraction of wood fuel is used to produce electricity in stand-alone condensing power plants. Both condensing power and cogeneration plantsare addressed below.

In order to evaluate electricity generation from wood fuels, the power generation cycles are considered first and the sizes available and thermal efficiencies of various options are analyzed and compared. The analysis addresses the three existing basic wood-fired power generation cycles: (1) steam-electric condensing power generation; (2) steam cycle cogeneration; and (3) modified gas turbine cogeneration. In all cases the combustion efficiencies and temperatures discussed in Chapter 5 are assumed here.

II. WOOD-FIRED CONDENSING POWER PLANTS

As Chapter 1 shows, wood is periodically used as a fuel for condensing power plants; such a plant is depicted schematically in Fig. 7-1. In considering this option, the basic technical parameters are scale and thermal efficiency of operation, which are interrelated factors.

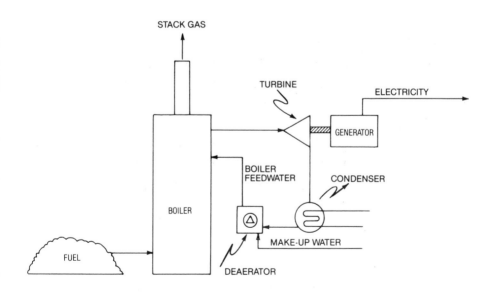

FIG. 7-1. A simplified schematic of a wood-fueled condensing power plant.

A. Scale Limits

One of the largest wood-fired boilers is the Longview No. 11 boiler of the Weyerhaeuser Co. It consumes approximately 90 tonnes (100 tons) of hog fuel/h at ~50 percent moisture (green), and produces 227×10^3 kg (500×10^3 lb) of 85 atm/783 K (1250 psig/950°F) steam (Jamison, 1979). Bethel et al. (1979) states that this boiler size may be near the practical limit of scale for wood-fueled facilities . This size limit is largely defined by materials-handling considerations including fuel storage and fuel transport into the firebox.

If one estimates net water rates for wood-fired condensing power plants at 3.6-4.5 kg steam/kWh (8-10 lb steam/kWh), and applies those rates to the Longview boiler, then the maximum wood-fired power plant is 50-62.5 MW. Specific water rates and heat rates (MJ/kWh) are a function of efficiency calculations. For purposes here, it is sufficient to note that the maximum scale of a wood-fired power plant is more than an order of magnitude smaller than a conventional coal-fired unit. This scale limitation seriously constrains the efficiencies attainable with wood-fired systems.

B. The Efficiency of Wood-Fired Power Plants

While condensing power plants are generally designed to the regenerative Rankine cycle, steam cycle efficiencies are limited by the Carnot cycle expression

$$\eta = (T_2 - T_1)/T_2 \tag{7-1}$$

where T_2 is the maximum (throttle) temperature (°K) of the steam and T_1 is the exhaust temperature (°K) of the steam.

Steam cycle efficiencies are constrained by scale, particularly as it influences throttle steam conditions. Wood-fired power plants are typically proposed in the 5-60 MW range. Table

TABLE I.   TYPICAL THROTTLE STEAM CONDITONS
AS A FUNCTION OF POWER PLANT SIZE

Power plant size (MW)	Throttle steam	
	atm/°K	psia/°F
5-10	41/672	600/750
10-20	58/714	850/825
20-40	85/783	1250/950
40-60	102/811	1500/1000

TABLE II.   THEORETICAL EFFICIENCIES OF WOOD-FIRED POWER PLANTS
AS A FUNCTION OF SCALE (PERCENT BASIS)

Scale (MW)	Theoretical efficiency	
	Wet wood	Dry wood
0-10	37	43
10-20	38	44
20-40	41[a]	47
>40	42	49

[a]Calculation: [(783 - 311)/783] x 0.68 x 100 = 41 percent.

I shows typical proposed throttle steam conditions as a function
of scale.   It is useful to compare this 41-102 atm (600-1500
psia) pressure range for wood-fired power plants to the 163-238
atm (2400-3500 psia) range typically found in modern coal-fired
power plants (Hill, 1977).   As a consequence of lower throttle
pressures, opportunities for reheat cycles virtually disappear
(Tillman, 1980).

Steam cycle efficiencies are also limited by exhaust conditions, typically 0.1-0.17 atm (3-5 in. Hg). Saturation temperatures range from 320 K (116°F) to 330 K (134°F). However, by permitting up to 10 percent condensation in the turbine, exhaust temperatures can be decreased to 311 K (100°F) to increase cycle efficiency (Hill, 1977). Exhaust conditions are not determined by scale, but by the quantity and quality of condensing water available.

Theoretical steam cycle efficiencies, then, are limited to 54-62 percent for the wood-fired plants. If one assumes that 68 percent is the maximum boiler efficiency when burning wet wood, and that 79 percent is the maximum efficiency when burning dry wood, then total theoretical system efficiency limits can be calculated. These are presented in Table II.

As a practical matter, attainable efficiencies do not even approach those shown in Table II. Friction losses occur in the turbine generator. Some electricity generated must be used internally and more importantly, entropy increases exceed theoretical minima. Furthermore, dry wood is not considered for most power plant designs, because it is too valuable in special (suspension burning) operations such as firing veneer dryers (Furman and Desmon, 1976) and possibly pulp mill lime kilns (K. Leppa, personal communication, October 22, 1980). Therefore, realistic power plant efficiencies are much lower.

Plant efficiences can be calculated by total material and energy balances, extending the techniques used in Chapter 5. Figure 7-2 presents such balances for a 35 MW system. For these designs the thermal efficiency value is 22 percent. The thermal efficiency for 5-10 MW systems typically is approximately 17 percent; for the 10-20 MW system, it is approximately 20 percent; and for 40-60 MW systems, it is approximately 24 percent.

More traditionally, these efficiency values are presented as heat rates and water rates. Heat rates are defined by

$$HR = (kWh_p - kWh_i)/F \qquad (7-2)$$

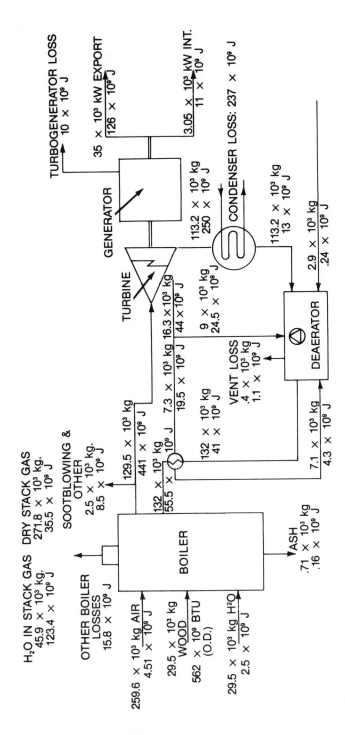

FIG. 7-2. Simplified material and energy balance for a 35 MW wood-fired power plant. (Basis of calculationL: 1 h.) The heat rate for this power plant is 14.9 MJ/kWh (from Tillman, 1980).

TABLE III. NET HEAT RATES FOR WOOD-FIRED POWER PLANTS
AS A FUNCTION OF SCALE

| | Heat rate | |
Plant scale (MW)	MJ/kWh	Btu/kWh
0-10	21.1	20,000
10-20	17.2	16,300
20-40	16.0	15,200
>40	15.0	14,200

TABLE IV. WATER RATES FOR CONDENSING TURBINE SYSTEMS
AS A FUNCTION OF THROTTLE PRESSURE

| Throttle pressure[a] | | Typical power plant water rate (net) | |
atm/°K	psi/°F	kg steam/kWh	lb steam/ kWh
41/672	600/750	4.4	9.8
58/714	850/825	3.9	8.7
85/783	1250/950	3.8	8.3
102/811	1500/1000	3.6	8.0

[a]Assumes 3 in. Hg exhaust.

where kWh_p is the total electricity production of a plant,
kWh_i the electricity consumption of the generating station
(e.g., for pumps, compressors, and pollution control), and F the
fuel consumed, expressed in joules or Btu. Heat rate is the
classic fuel efficiency measure. Heat rates are presented in
Table III as a function of scale.

Water rates are defined as steam (kg or lb) raised in the boiler per kWh produced. They are largely a measure of the efficiency of steam usage. Water rates for wood-fired systems are presented in Table IV as a function of steam condition differentials.

It is revealing to compare water rates of wood-fired plants with modern coal-fired power plants. These coal plants typically have water rates approaching 2.7 kg steam/kWh (6.0 lb/kWh) and net heat rates of 10 MJ/kWh (9500 Btu/kWh). Gains are available from using extensive feedwater heating, reheat cycles, and high pressure (>163 atm) high-temperature (840 K) steam. Additionally, the coal combustion process enjoys a high thermal efficiency (e.g., 85 percent) relative to wood fuel.

From a technical perspective, then, wood-fired power plants are limited to sizes of 5-60 MW, and heat rates of 21.1-15 MJ/kWh. These are serious limitations for condensing power plants.

III. STEAM CYCLE COGENERATION

Cogeneration is the sequential production of electricity and useful heat at the same facility. While all fuels can be used in this mode of power generation, wood is a prime candidate for use in such systems. Two basic cogeneration cycles exist: topping and bottoming. Furthermore, topping cycles can be divided into steam turbine, gas turbine, and diesel topping cycles. Of these, the steam turbine topping cycle, depicted schematically in Fig. 7-3, is most appropriate for wood, given currently available technology. In this cycle, high-pressure steam is expanded through a turbine and exhausted at industrial process conditions.

Normally, cogeneration systems are installed by manufacturers with large, stable needs for process steam. Such manu-

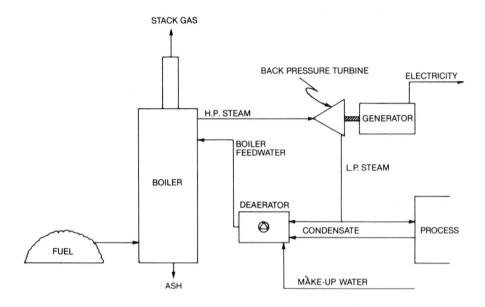

FIG. 7-3. Simplified schematic of a wood-fired cogeneration
system.

facturers include the forest products industry, which consumes
saturated steam at such pressures as 27.2 atm (400 psia), 10.2
atm (150 psia), and 3.4 atm (50 psia) in lumber drying, pulp
digesting, and other operations. Mill steam demands range from
45×10^3 kg/h (100×10^3 lb/h) to 450×10^3 kg/h (10^6
lb/h) and more. Cogeneration is accomplished by noncondensing
back-pressure or automatic extraction turbine generators inserted between the boiler and the industrial process (see Fig.
7-3). They have thermodynamic advantages over both process
steam boilers and condensing power plants.

Process steam boilers typically have flame temperatures in
the 1480-1750 K range (2200-2700°F) when burning wet wood fuel.
Process steam at 3.4-10.2 atm of pressure, however, has a temperature range of 411-502 K (281-445°F). Consequently, consi-

derable work potential exists in the differential between the flame and the process steam temperatures. Much of this potential can be captured by raising steam to 672-783 K (750-950°F) and increasing the pressure accordingly. In reducing the temperature differential between the combustion condition and the product steam condition, cogeneration enjoys thermodynamic advantages over merely raising process steam. This concept is thoroughly explored in Stobaugh and Yergin (1980) and Gyftopoulos et al. (1974).

The advantage of cogeneration over condensing power plants stems from the former's ability to capture the heat of vaporization in the steam. As Fig. 7-2 shows, condenser energy losses can be about double the energy content of the salable power. Because industrial processes can capture the 2.3 MJ/kg (970 Btu/lb) of heat available in the phase change from steam to water, cogeneration power plants need not absorb this as a loss attributable to power production.

Because cogeneration systems typically are installed by manufacturers, the steam demands of the manufacturer typically size the equipment. Thus, it is important to examine typical power-to-heat ratios and throttle steam conditions as a function of system scale. Power-to-heat ratios are determined by the equipment itself, which is constrained by manufacturing conditions. In this case, it is essential to note that a lower water rate does not necessarily connote a lower heat rate: it is more a determinant of power/steam ratios. This separation of water rate from heat rate as a measure of efficiency, results from the simultaneous production of heat and power.

The influence of throttle and exhaust steam conditions on power/heat ratios can be seen from Fig. 7-4 (Kovacik, 1980). The turbine absorbs more heat energy from the steam (before sending it to process) as one moves upward and to the left in Fig. 7-4. While the requirements of the manufacturing process generally dictate exhaust steam conditions, the quantity of

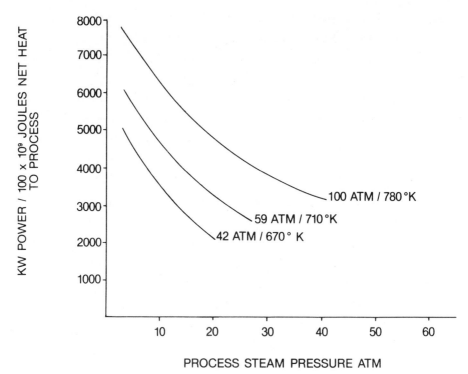

FIG. 7-4. The influence of steam conditions on electricity generation in wood-fired cogeneration systems. Throttle steam conditions are shown on the figure. Process steam pressure is shown on the x axis (from Kovacik, 1980).

steam required generally estabishes the throttle steam conditions. Table V shows normal throttle conditions as a function of boiler size. Table VI shows the influence of turbine size/-capacity on water rate, assuming a throttle steam condition of 85 atm/783 K (1250 psia/950°F) to 41 atm/672 K (600 psia/750°F) and exhaust conditions of 10.2 atm (150 psia). Such values exist for all back pressure and extraction turbines used in co-generation installations.

It should be noted that low water rates are highly desirable, not for efficiency reasons, but for economic reasons. The

TABLE V. THROTTLE STEAM CONDITIONS
AS A FUNCTION OF BOILER SIZE[a]

Boiler size		Mininum		Maximum	
10^3 kg/h	(10^3 lb/hr)	atm/°K	(psia/°F)	atm/°K	(psia/°F)
45	(100)	27/616	(400/650)	58/714	(850/825)
91	(200)	27/616	(400/650)	85/755	(1250/900)
136	(300)	41/672	(600/750)	99/783	(1450/950)
227	(500)	58/714	(850/825)	99/783	(1450/950)

[a]From Kovacik (1980).

value of a joule in the form of electricity is substantially more valuable than a joule in the form of steam. It is also significant that water rates for cogeneration are higher than those associated with condensing power by a factor of 2 to 4. Further, large exclusively wood-fueled cogeneration systems are approximately 30 MW (this is exclusive of systems based on black liquor recovery boilers). Figures 7-5 to 7-7 depict a 30 MW system, the Longview No. 11 boiler of Weyerhaeuser.

Thermal efficiency for cogeneration is generally measured as heat-chargeable-to-power. Heat-chargeable-to-power is a measure comparable to heat rate for a central station power plant. Its calculation is more complex, however, as shown below:

$$HCP = (kWh_p - kWh_b)/(F_T - F_S) \qquad (7-3)$$

where kWh_p is total power production, kWh_b the power consumed in operating the boiler and turbine-generator, F_T the total fuel consumed (J) and F_S the fuel that would be consumed if the plant raised only process steam. The value of $F_T - F_S$ represents the fuel required to increase the enthalpy of the steam from process levels to turbine throttle conditions.

TABLE VI. WATER RATES FOR BACKPRESSURE TURBINES
AS A FUNCTION OF SCALE AND THROTTLE PRESSURE[a,b]

Turbine generator capacity	Throttle steam conditions					
	85 atm/783 K (1250 psia/950°F)		58 atm/714 K (850 psia/825°F)		41 atm/672 K (600 psia/750°F)	
(MW)	kg/kWh	(lb/kWh)	kg/kWh	(lb/kWh)	kg/kWh	(lb/kWh)
5	9.4	(20.7)	11.7	(25.8)	15.4	(34.0)
10	8.7	(19.2)	11.0	(24.3)	25.9	(32.8)
15	8.3	(18.4)	10.7	(23.6)	14.3	(31.6)
20	8.2	(18.1)	10.6	(23.4)	14.0	(30.9)
25	8.1	(17.9)	10.5	(23.2)	13.9	(30.6)
Theoretical	6.5	(14.4)	8.4	(18.6)	11.1	(24.5)

[a]From C. Bruya (personal communication, May 2, 1979).
[b]Assumes 10.2 atm or 150 psig exhaust steam.

FIG. 7-5. Fuel supply to the Weyerhaeuser Longview No. 11
boiler. This boiler consumes approximately 90 tonnes/h of hog
fuel. Note the similarity on particle size variation between
this figure and Fig. 3-4. Also note the extensive fuel-handling
equipment requirement associated with wood-fueled systems (from
Jamison, 1979).

FIG. 7-6. The Longview No. 11 boiler. This wood-fired installation produces 227,000 kg or 227 metric tons of high-
pressure steam for the production of electricity by cogeneration
and for subsequent use by process.

Heat-chargeable-to-power is depicted in Fig. 7-8 for a 5 MW
system. This 5 MW system is close to the lower size limit for a
viable system. Typical values are in the vicinity of 5.6 to 6.3
MJ/kWh (5300 to 6000 Btu/kWh). This compares to cogeneration
heat rates of 4.4 MJ/kWh (4200 Btu/kWh) for coal-fired systems.

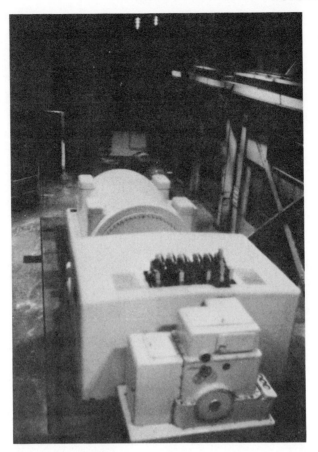

FIG. 7-7. The 30 MW back-pressure turbine associated with the Longview No. 11 boiler.

The efficiency gains available from cogeneration can be measured by the following ratio:

$$G_{CG} = HR_{CP}/HR_{CG} \qquad (7-4)$$

where G_{CG} is the ratio measure of gains from cogeneration, HR_{CP} the heat rate for condensing power, and HR_{CG} the heat rate for cogeneration. The wood G_{CG} ratio is in the range 2.4-3.3. The coal value range is 2.0-2.1. Gains are higher for wood, vis-a-vis coal, because the turbine-generator and process

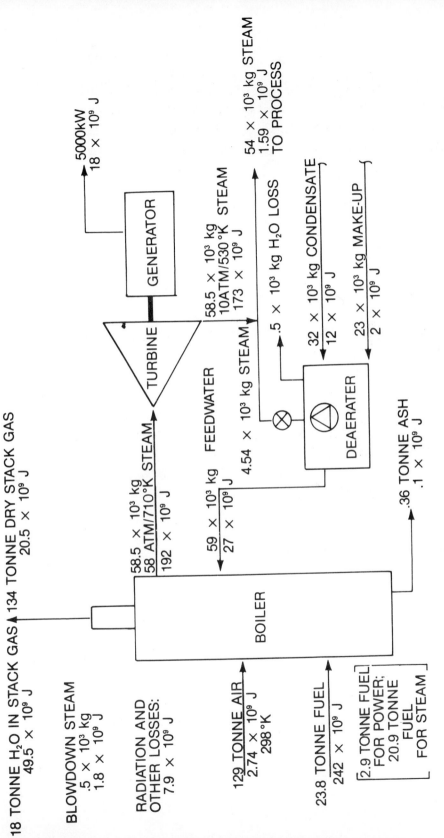

FIG. 7-8. Simplified material and energy balance for a 58.5 metric ton steam, 5 MW cogeneration system (from Tillman and Jamison, 1981).

are indifferent to the fuel used to raise the steam. Thus, while wood is at a competitive technical disadvantage in condensing power applications, that technical disadvantage is significantly reduced when steam cycle cogeneration is employed.

IV. MODIFIED GAS TURBINE COGENERATION

Steam cycle power plants require extensive feedwater treatment systems and have significant temperature limitations on the T_2 value of the working fluid [see Eq. (7-1)]. Further, heat must be applied to the working fluid to accomplish the water-to-steam phase change. A modified gas turbine cycle proposed by Hagen and Berg (1976) addresses these problems for wood fuel. The basic cycle also appears in Fryling (1966), applied to coal. The original Hagen and Berg proposed is for a regenerative cycle. However, it can be modified to be an electricity/direct heat cycle, which might be employed in a large plywood mill. This option is shown in Fig. 7-9.

Alternatively, one can visualize a wide array of operations from Fig. 7-9 including simple cycle, regenerative cycle, combined cycle, and a steam-producing cogeneration or combined system. The principal limitation in this system is the ash softening temperature of wood fuel, which may be as low as 1370 K (2000°F).

This system exhibits promise as a means for expanding cogeneration applications of wood fuel and possibly altering the power/process heat ratio. It awaits successful development of a high-temperature ceramic heat exchanger capable of withstanding possible fouling from wood ash (Bethel et al., 1979). Hague International is addressing this problem with a silicon carbide heat exchanger capable of withstanding temperatures up to 1600 K (2400°F) according to Hagen and Berg (1976).

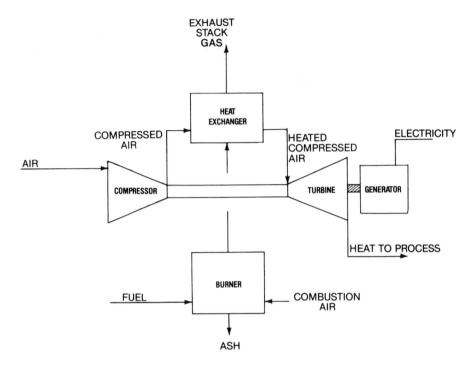

FIG. 7-9. Schematic representation of a combustion turbine cogeneration system as proposed by Hagen and Berg (1976). In this system, the significant losses are not stack losses, but are losses of energy from the turbine associated with compression of the incoming air.

The scale of operation generally discussed by proponents of this system ranges from 1.5 to 5 MW. Hagen and Berg (1976) have calculated a heat balance for the 1.5 MW system, which shows that the steam-to-power ratio for this system (comparable to the water rate) is 7.5 kg/kWh (16.6 lb/kWh). The overall thermal efficiency is 58 percent. This is in contrast to the water rate of 11.6 kg/kWh and the overall thermal efficiency of 65 percent shown in Fig. 7-8 for a small steam topping cycle. Heat chargeable to power is 12 MJ/kW, almost double that of the steam turbine cycle. In this regard, it is useful to note that parasitic compression power requirements are 13.3 percent of the total fuel requirement. Those losses are higher than dry stack gas losses.

The modified combustion turbine cycle, then, has potential in small-scale cogeneration applications, particularly where high power-to-steam ratios are desired. It capitalizes on advantages of combustion turbine technology that are not available to the steam turbine cycles. Thus, its emergence in the marketplace can provide yet another opportunity for wood fuels.

V. COMPARISON OF POWER GENERATION CYCLES

Overall technical comparisons between systems are presented in Table VII. From this table it is clear that the most efficient system, both in terms of heat rate and overall efficiency,

TABLE VII. COMPARISON OF POWER GENERATION CYCLES

Comparison parameter	Cycle		
	Steam condensing	Steam cogeneration	Combustion turbine cogeneration
Scale of operations (MW)	5-60	5-30	1.5-5
Representative water rate (kg/kWh)	4.1	11.1	7.5
Representative heat rate (MJ/kWh)	15.8	6.3	12.1
Overall system efficiency (percent)	22	65	58

is the steam turbine topping cycle. This system requires the largest boiler, however, as indicated by the water rate. The combustion turbine cycle is best suited to small-scale systems required where overall process energy requirements warrant its application. Condensing cycles have the lowest thermal efficiency and the lowest water rates. They are best suited to applications where there is an abundance of fuel and where the need for an efficient system is not great.

There are other electricity generation cycles for wood fuel including the production of lowor medium-Btu gas and the use of that gas either in a combustion turbine or in a diesel generator. These are novel cycles that do not use wood combustion and are therefore not considered here. In general, however, it can be observed that, while wood fuel is less efficient than fossil fuel, the efficiency differential can be minimized by application of cogeneration systems.

REFERENCES

Bethel, J. S., et al. (1979). "Energy from Wood." A Report to the Office of Technology Assessment, Congress of the United States. College of Forest Resources, Univ. of Washington, Seattle, Washington.

Fryling, G. R., ed. (1966). "Combustion Engineering." Combustion Engineering, Inc., New York.

Furman, L. H., and Desmon, L. G. (1976). Wood residue for veneer drying: A case history, Forest Prod. J., September.

Gyftopoulos, E. P., Lazaridis, L. J., and Widmer, T. F. (1974). "Potential Fuel Effectiveness in Industry." Ballinger Publ. Co., Cambridge, Massachusetts.

Hagen, K. G., and Berg, C.A. (1976). Wood residue fired gas turbine cycle, in "Energy and the Wood Products Industry." Proc. Forest Prod. Res. Soc. Mtg., 123-135. Madison, Wisconsin.

Hill, P. G. (1977). "Power Generation: Resources, Hazards, Technology, and Costs." MIT Press, Cambridge, Massachusetts.

Jamison, R. L. (1979). Wood fuel use in the forest products industry, in "Progress in Biomass Conversion," Vol. 1 (K. V. Sarkanen and D. A. Tillman, eds.), pp. 27-52. Academic Press, New York.

Kovacik, J. M. (1980). Design considerations for large industrial cogeneration systems, paper presented at the Oregon Governor's Cogeneration Workshop, Portland, Oregon, April 20-21.

Stobaugh, R., and Yergin, D. (1979). "Energy Future: Report of the Energy Project at the Harvard Business School." Random House, New York.

Tillman, D. A. (1980). Fuels from waste, in "Kirk-Othmer Encyclopedia of Chemical Technology," Vol. 2, 3rd ed. Wiley, New York.

Tillman, D. A., and Jamison, R. L. (1981). A review of cogeneration with wood fuel, Fuel Proc. Technol., in press.

CHAPTER 8

ECONOMIC AND FINANCIAL ISSUES
ASSOCIATED WITH WOOD COMBUSTION

I. INTRODUCTION

Technical considerations related to fuel supply, materials
handling, combustion and heat release, pollution control, and
power generation discussed in earlier chapters can be integrated
through economics techniques. Economics translates technical
and other considerations into costs enabling effective assessment of wood energy projects. Economics penalizes inefficient
systems and, through the costs of pollution control, interna-
lizes some of the costs of pollution control and mitigation of
processes that emit substantial quantities of effluents. Economics provides a means for dealing with concerns such as technical risk and innovation.

The economics of wood-fueled energy systems merits special
consideration because to a large degree it dictates how wood
combustion will contribute to the U.S. economy, particularly in
the area of electricity supply; which fuels will be used (e.g.,

residues, fuel farm materials); and which systems can be employed (e.g., condensing power, cogeneration). The applicabi-
lity of supply and utilization systems, as well as factors related to fuel cost are for a range of systems. The systems considered vary and include representative condensing power and
cogeneration options.

A. Analysis Framework

This section is based on the premise that the primary objec-
tive of any corporation is to maximize the wealth of its stockholders; investments are made to accomplish that purpose. Devi-
ation from it can result in a loss of competitive position and,
ultimately, bankrupcy.

Because the goal of any corporation is wealth maximization,
investments are made that provide a positive net present value
(NPV) to the corporation. NPV is defined as

$$NPV = \sum_{t=1}^{n} [NATCF/(1 + i)^t] - (I - ITC) \tag{8-1}$$

where t is the year with respect to the investment (considered
to be made in year 0), n the final year of the project life,
NATCF the net after tax cash flow, i the cost of money or discount rate; I, investment cost, and ITC, investment tax credits.

Equation (8-1) defines the critical considerations in assessing investments: (1) the period of the investment; (2) the
cash flows, which are not equal to profits; (3) the discount
rates; and (4) the investment tax credits available. Equation
8-1 can be written in any number of ways (see Bierman and Smidt,
1971; Theusen et al., 1977; Schall and Haley, 1977), all of
which achieve the same result.

The key variables can be rearranged as nonownership- and ownership-related terms. The nonownership variables are I and NATCF, with the latter calculated as follows:

$$NATCF_t = [R - (O \text{ and } M_t + D_t)] \times (1 - TR) + D_t \qquad (8\text{-}2)$$

where R is the revenue stream, O and M the annual operating and maintenance costs in year t, D_t the depreciation term in year t, and TR the marginal tax rate.

The most critical nonownership variables are the investment costs and O and M costs, which are significant largely because of fuel cost. The investment cost includes purchase and erection of the boiler, turbine-generator, and related equipment. The fuel cost is determined by quantity of fuel demanded. Both are examined in Section II,A.

Ownership-related variables include available investment tax credits, project lives (and hence depreciation periods), and appropriate discount rates. Investment tax credits, project lives, and depreciation are accounting functions. These functions, along with calculations of the discount rate, are reviewed in Section II,C. In Section II,D, project economics are reviewed.

II. NONOWNERSHIP VARIABLES

For electricity-producing systems, the critical variables are capital investment and fuel costs. These variables are addressed below.

A. Capital Costs

Only those capital costs chargeable to power are analyzed in economic evaluations of wood energy systems. Furthermore, the capital investment chargeable to power is largely a function of

TABLE I.   TYPICAL CAPITAL COSTS FOR A 5 MW COGENERATION FACILITY[a]

	Cost allocation	
	Steam	Power
Cost category	raising	generation
Direct costs		
site preparation	0.1	--
fuel handling	1.5	--
boiler and related		
(including emission controls)	3.6	
turbine-generator and accessories	--	1.7[b]
Subtotal	5.2	1.7
Engineering at 15 percent	0.8	0.3
Commissioning and contingencies		
at 20 percent	1.0	0.4
Working capital at 5 percent	0.3	0.1
Total	7.3	2.5
Installed cost/kW		500

[a]Values in 1980 dollars x 10^6.

[b]Includes incremental costs of pressure parts.

the type of system installed. For condensing power systems, the electricity must bear the entire capital investment. Tables I and II show capital costs for representative cogeneration and condensing power systems, updated to 1980 dollars.

Two important conclusions can be drawn from Tables I and II. First, the facts that $1.5 million of the $5.2 million spent on direct costs of steam raising is required for fuel handling and an additional $3.6 million is spent on boiler and related facilities suggest that wood is difficult to handle and burn efficiently.

TABLE II. TYPICAL CAPITAL COSTS
FOR A WOOD-FIRED CONDENSING POWER PLANT[a]

	Power plant size	
Cost category	5 MW	25 MW
Direct Costs		
boiler and related	6.1	17.4
turbine-generator and related	2.4	4.3
condenser and cooling towers	0.4	1.1
turbine-generator and accessories	--	1.9
auxiliary equipment	0.1	0.6
substation and switchgear	0.4	0.7
Subtotal	9.4	24.1
Indirect costs		
Engineering at 15 percent	1.4	3.6
Commissioning and contingencies		
at 20 percent	1.0	4.8
Working capital at 5 percent	0.5	1.2
Total	13.2	33.7
Cost/installed kW	2640	1350

[a]Values in 1980 dollars x 10^6.

Second, it can be seen that the investment for power generation is much lower for cogeneration than condensing power systems. For example, as shown in Tables I and II, a 5 MW condensing power plant has a capital cost 5.3 times that of the incremental cost of a similarly sized cogeneration plant. For most
plants, data presented by Tillman and Jamison (1981) show an
incremental investment cost of $500/kW for wood-fired cogenera-
tion systems. The capital costs for condensing plants are
approximately 2.7 times those of cogeneration plants.

B. Wood Fuel Costs

The cost of wood fuel at the conversion facilities is most
easily understood in terms of two microeconomic principles: (1)
the marginal costs curve and (2) the opportunity cost principle. The first concept aids in understanding the cost of residual fuels. The second concept is most relevant in determining
the cost of fuel from silvicultural energy farms.

As applied to wood fuels, the marginal cost concept shows
that, in the short-range time frame, the production of increasing quantities of fuel requires the allocation of resources in
a less efficient manner. Hence, production costs escalate at
the margin as quantities made available increase.

The opportunity cost principle states that the cost of a
given commodity is equal to the value of the opportunities foregone. If a landowner sells all of his stumpage as fuel wood,
the cost of the fuel is equal to the opportunity foregone. In
the case of fuel wood the increased revenue that would have re-
sulted from selling the stumpage as sawlogs or otherwise to
bring the highest price.

1. The Cost of Wood Residual Fuels
The price of any commodity is determined by the intersection
of the supply and demand curves. As discussed earlier, the supply curve, or marginal cost curve, for wood fuels is a local
phenomenon as determined by the costs of accumulating such fuels
locally. Such costs are largely associated with incremental expenditures made for handling and storage systems and expenditures not made for disposal of such wastes when the fuel materials have no other use. When residues have other uses (e.g.,
particle boards plants or pulp mills), the fuel price is the
cost of not using such residues in alternative operations.

Forest residue fuel costs are a function of the expenditures
made for gathering, processing, and transporting such materials

to the mill. These costs are, in turn, a function of residue piece size, forest terrain, and distance from the mill. From these costs, current disposal charges can be subtracted since there is little current use for forest residues and their use eliminates a waste disposal problem. The exact shape and configuration of this cost function varies according to regional considerations.

In the Southeast and Midsouth, terrain is generally less rugged than in the West. This permits use of mechanized harvesting systems such as feller-bunchers, wheel skidders, and whole-tree chippers. While such systems can be deployed to varying degrees throughout the country, they are appropriate in the South. Thus, piece size becomes the dominant variable in this region.

In the Southeast and the Midsouth, there are millions of acres of mixed hardwood-softwood stands. For example, there are 4.8×10^6 ha (11.9×10^6 acres) of oak/pine forest in the southeast, of which the forest industry owns 0.7×10^6 ha (1.8×10^6 acres). In the Midsouth there are 6.5×10^6 ha (16.3×10^6 acres) of oakpine forest, of which the forest industry owns 1.7×10^6 ha (4.2×10^6 acres) (Forest Industries Council, 1980). These stands produce higher than average logging residue volumes in harvesting due to the different merchandizing standards for hardwoods, and also due to higher levels of defect in the deciduous species.

In the Southeast and Midsouth, there are also numerous stands composed primarily of undesirable hardwood species. These stands can only achieve economic productivity if the hardwoods are removed and the land is replanted in pine. The Forest Industries Council (1980) identifies 1.3×10^6 ha (3.3×10^6 acres) of such lands owned by forest products companies, with 1.0×10^6 ha (2.5×10^6 acres) being in the Midsouth. There are also significant opportunities for stand conversions outside the land owned by the forest products industry. The forest industry lands will, however, probably be converted to softwoods more rapidly than the 4.8 x 10^6 ha (11.8 x 10^6 acres) in other ownerships that offer the same potential.

Mill and logging residuals are classified, economically, as flows; and the annual quantities produced are a function of the demand for structural and fiber products. Stand conversion materials, however, are considered as stocks, and they are converted into flows at given rates (e.g., 5-10 percent of the material available). A 5 percent rate is assumed in silvicultural residues.

Given such flows, a marginal cost curve has been calculated for one location in the Midsouth. That curve is shown in Fig. 8-1. It may be considered as representative of supply functions in that region. Similar curves have been synthesized for residues in the Pacific Northwest. Generally, the intercept is about $2.00/GJ, and the slope rapidly exceeds unity as piece size and terrain conditions play significant roles as independent variables in such curves (Bethel et al., 1979; Envirosphere, 1980).

What becomes clear from these residue supply systems, and the attendant costs, is that wood fuels become increasingly expensive as one moves from the mill to the forest. More factors of production must be allocated to their collection, processing, and transportation; the consequence is a rising marginal cost curve.

2.   The Cost of Producing Wood Fuels on Silvicultural Fuel Farms

The true cost of wood from silvicultural fuel farms is best understood in terms of the opportunity cost of growing and harvesting such materials. The opportunity cost can be calculated as the net present value (NPV) foregone by not raising commercial timber species such as loblolly pine in the South or Douglas fir in the Pacific Northwest. In order to calculate the

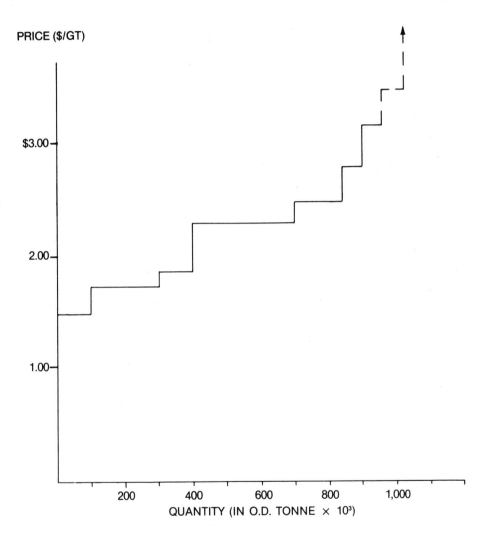

FIG. 8-1. Marginal cost curve for wood fuel in the Oklahoma area. Note that as the quantities of fuel required per year are increased, the price of fuel to all wood consumers increases (from Jamison et al., 1978).

cost of fuel farm fuels, a stumpage value must be ascribed to the wood grown. The sum of the stumpage charges, discounted over time, must be added to the production cost such as harvesting and transportation.

This analysis has been conducted for the Southeast by Tillman (1980a). Assuming site index 60 (25 yr basis) and various regimes, some 210-300 M^3/ha (30-49 CCF/acre) of timber can be grown [see Fender (1968) for lower yields, Feduccia et al. (1979) for higher yields] without fertilization. At optimal application levels, fertilization can add 31 m^3/kg (2 ft^3/- lb) of nitrogen fertilizer. The use of 136 kg of nitrogen over a 30 year rotation then can add 42 m^3/ha of roundwood.

Given these values, stumpage cost are $2.75/GJ ($2.90/10^6 Btu) for the 210 m^3/ha scenario and $4.80/GJ if 330 m^3/ha are expected over a 30 yr period. Both values assume a 5.4 percent annual increase in stumpage values (U.S. Forest Service, 1979). Production costs of a $1.25/GJ must be added to these stumpage values if conventional mechanized harvesting is used. Thus, the delivered cost of fuel farm energy is in the $5.00- 6.05/GJ ($5.30-6.40/10^6 Btu) range, in year 1 of production.

Costs for fuel farm materials grown in the Pacific Northwest are expected to be higher than those associated with fuels raised in the Southeast for several reasons: (1) rotations are longer for red alder than for American sycamore; (2) the commercial species, Douglas fir, is more valuable than its counterpart, loblolly pine (U.S. Forest Service, 1979); and (3) fuel farm production costs are higher (Tillman, 1980a).

In both regions wood grown specifically for fuel purposes is expected to cost substantially more than residues gathered up and sold or used for energy purposes. This point can be seen by comparing the $4.00-6.05/GJ for the Southeast to the values shown in Fig. 8-1. Residues have the distinct advantage of not carrying the value of productive land as part of their cost.

3. Application of Fuel Costs to Electricity Prices

Fuel costs are the primary operating cost in the generation of power from wood burning, either by cogeneration or condensing power. The price of wood fuel is set by the intersection of the supply and demand curves as discussed earlier.

In cogeneration, however, an additional consideration is involved. If one assumes a small system that does not change the price of wood fuel, then fuel costs are calculated as follows:

$$C_{CGF} = Q_{CGF} \times P_F \tag{8-3}$$

where C_{CGF} is the annual cost of fuel for cogeneration, Q_{CGF} the quantity of fuel used for power generation (calculated from the heat chargeable to power) and P_F the price of fuel determined by the marginal cost curve.

If the installation of cogeneration causes a rise in fuel prices, that rise affects the process-steam-raising function of the boiler as well as the power generation function. Such a case might occur in the Midsouth if annual fuel requirements rose from 90×10^3 to 120×10^3 t. In such a case, fuel costs chargeable to power could be calculated as follows:

$$C_{CGF} = (Q_{CGF} \times P_F) + Q_{SF} \times (P_F - P_{SF}) \tag{8-4}$$

where Q_{SF} is the quantity of fuel used for process steam raising and P_{SF} the price of fuel incurred for steam raising only.

Using the example previously stated, Q_{CGS} is 30×10^3 t/yr. From Fig. 8-1, it has a cost of $1.50/GJ. The $Q_{CGF} \times P_F$ is calculated as follows:

30×10^3 t/y x 18 GJ/t x $1.50/GJ = $810,000/yr

The adjustment for increasing the cost of fuel for steam raising then is as follows:

90×10^3/y x 18 GJ/t x ($1.50 - $1.20)/GJ = $486,000/yr

The total fuel costs chargeable to power are $1,296,000--or $43/t ($2.40/GJ). A steeply sloping marginal cost curve can seriously influence the economics of cogeneration or other wood-burning projects.

III. OWNERSHIP-RELATED VARIABLES

Two classes of ownership variables are considered here: accounting variables and capital cost variables.

A. Accounting Variables

Accounting variables include depreciation period and investment tax credits available. These are presented in Table III for wood fueled facilities. Analysis of Table III reveals that

TABLE III. DEPRECIATION SCHEDULES AND INVESTMENT TAX CREDITS AVAILABLE FOR VARIOUS FACILITIES[a]

Facility	Depreciation range (yr)	Investment tax credit (percent)
Sawmill power plant	8-12	20[b]
Pulp mill power plant	13-19	20[b]
Utility-owned condensing power plant	22.5-33.5	10

[a]From Internal Revenue Service (1977) and National Energy Act.

[b]Includes energy tax credits.

more accelerated depreciation schedules are available for sawmills and pulpmills for the utility sector. Similarly, higher
investment tax credits are available to the forest products industries.

B. Discount Rates

Discount rates are generally calculated on an industry-wide
basis, and therefore the weighted average cost of capital (WACC)
technique is appropriate for their estimation.

The WACC calculations are based upon the Modigliana and
Miller model as described by Haley and Schall (1979). The general equation is

$$K_V + (1 - \theta)k_S + \theta k_B \qquad (8-5)$$

where kV is the cost of capital, θ the proportion of capital
supplied by debt, k_S the cost of equity capital, and k_B the
cost of debt capital. The model argues that k_B is lower than
k_S; however, increasing θ to decrease k_V results in higher
expectations of both bond holders (increasing marginal costs of
borrowed money) and stockholders (increasing as a result of
business risk assessment). Thus k_V will not, in fact, decline. This approach applies to normal business conditions and
a stable debt/equity ratio; and it must be altered if unusual
events or investments occur.

The Modigliani-Miller model approximates the nominal discount rate, whose structure is

$$DR = I + P_M + P_R \qquad (8-6)$$

The structure of discount rates is generally considered to
have three additive components: (1) expected inflation (I), (2)
premium for early availability of funds (P_M), and (3) premium
for risk (P_R). The real discount rate, however, has only the
second and third terms--eliminating inflation expectations.

TABLE IV.   CAPITAL STRUCTURES AND TAX RATES
FOR THREE FOREST PRODUCTS FIRMS

		Firm	
Capital structure	Boise Cascade	Weyerhaeuser	Georgia Pacific
Proportion as debt	0.187	0.194	0.177
Proportion as preferred stock	0.009	0.034	0.009
Proportion as common stock	0.804	0.772	0.814
Effective tax rate	0.350	0.304	0.430

Thus, the calculation of real rates hinges on estimating inflation expectations.

1.   The Cost of Capital for Forest Industry Facilities
In order to finance the cogeneration facilities, the
weighted average of cost of capital for the forest products industry is calculated.   Three representative companies are employed with the values of a median firm--the Weyerhaeuser Co.--
chosen for subsequent analysis.   Data for these analyses are
taken from annual reports of Georgia Pacific, Boise Cascade, and
Weyerhaeuser Companies; from the Standard and Poor's Bond Guide
(1979); and from the Wall Street Journal, August 28, 1979.
Table IV shows the estimated capital structure of the three
firms analyzed, along with tax rates applicable.   From this
table, three salient points can be drawn: (1) the tax rate is
heavily influenced by land and timber ownership; (2) the dominant capital proportion comes from common stock; and (3) because
common stock is dominant, corporate growth rates play an ex-

TABLE V. VALUE LINE GROWTH MEASURES[a,b]

| | Years in | Firm | | |
| | trend | Boise | | Georgia |
Growth measure	analysis	Cascade	Weyerhaeuser	Pacific
Cash flow	10	6.5	15.0	12.0
	5	24.0	11.5	10.0
Earnings	10	8.5	14.5	13.5
	5	38.0	10.5	13.0
Dividends	10	16.5	9.0	14.0
	5	44.0	14.0	13.5
Est. earnings growth	5	14.5	13.5	11.0
Earnings/share,				
10 yr	10	8.4	13.7	13.3

[a]From Value Line (1979).
[b]Percent basis.

tremely important role in determining the discount rate.[*]
Thus, Table V is posited, with numerous value line (Value Line, 1979) growth measures for the forest products firms analyzed. The growth measure employed is the 10 yr earnings per share trend. Based upon these data, discount rate estimates for the firms are presented in Table VI.

The three components of the forest industry discount rate chosen are: $I = 6.5$, $P_M = 3.0$, and $P_R = 4.8$ percent. The

---

[*]The cost of common equity capital is determined by $D_e/P + G_e = K_{ceq}$ where D_e is the expected dividend, P the price, G_e the expected growth, and K_{ceq} the capital cost, common equity.

TABLE VI. WEIGHTED AVERAGE COST OF CAPITAL
FOR SELECTED FOREST INDUSTRY FIRMS[a]

	Firm		
Capital instrument	Boise Cascade	Weyerhaeuser	Georgia Pacific
Debt	1.8	1.2	1.0
Preferred	Negl.[b]	0.2	0.1
Common	9.2	12.9	13.9
Total	11.0	14.3	15.0

[a]In percent.
[b]Calculated at 0.01 percent; this stock is not being traded.

real (inflation-free) rate to be employed is 7.8 or 8 percent.
The 6.5 percent inflation term reappears as the deflator rate.
The sum of these terms equals 14.3 percent.

There is a precedent for using a nominal discount rate of 14
percent, equal to a real discount rate of 8 percent. For the
30-yr period 1946-1975, corporate stock investments yielded
approximately 11 percent while inflation, as measured by either
the Consumer Price Index or the Producer Price Index, averaged
approximately 3 percent (Alchian and Allen, 1977; U.S. Bureau of
Census, 1975, 1978).

2. Cost of Capital, Electric Utilities
An examination of privately held utilities and utility
holding companies shows that the WACC is not particularly sensitive to region. Thus, five utility corporations were employed
in order to obtain a cost of capital: (1) Public Service Co. of
New Hampshire, (2) New England Power Co., (3) The Southern Co.,

TABLE VII. THE CAPITAL STRUCTURE OF FIVE UTILITIES

Utility	Debt proportion	Preferred stock proportion	Common stock proportion	Total
Public Service Co. of New Hampshire	0.521	0.074	0.405	1.00
Washington Water Power	0.506	0.049	0.445	1.00
Puget Sound Power and Light	0.526	0.092	0.382	1.00
The Southern Co.	0.569	0.121	0.310	1.00
New England Power Co.	0.664	0.086	0.250	1.00

TABLE VIII. THE CAPITAL COST OF SELECTED UTILITIES

Utility	Nominal cost (percent)	Real cost (percent)	Effective tax rate
Public Service Co. of New Hampshire	12.9	6.4	0.23
Washington Water Power	12.8	6.3	0.22
Puget Sound Power and Light	11.4	4.9	0.18
The Southern Co.	10.7	4.2	0.36
New England Power Co.	10.3	3.8	0.20

(4) Washington Water Power, and (5) Puget Sound Power and Light. All are publicly held, and represent the regions considered. Other utilities considered include the Middle South Utilities and Central Vermont Public Service.

Table VII shows the capital structure of the five utilities detailed. Table VIII provides estimates of the cost of capital for each firm. The result is a nominal cost of capital of 11.5-12 percent.

It is significant to note that the risk level for utilities is perceived to be substantially below the risk for the forest products industry in the financial market. This is due to the guaranteed return on capital associated with utilities. The average utility return on common stock was 9.1 percent from 1946 to 1971, compared to 12.8 percent for industrial common stocks (Hass et al., 1974), a differential attributed to risk assessment. However, the data suggest that the discount rate for utilities is, indeed, rapidly rising. Bond yields have almost doubled since 1963, and since 1969 there has been a substantial derating of utility bonds by both Moody's and Standard and Poor's (Hass et al., 1974). Standard and Poor's A rated bonds averaged 4.39 percent yield to maturity in 1963; in August, 1979 Standard and Poor's shows a 9.3-9.8 percent range (Standard and Poor's, 1979). These yields are no longer substantially different from A rated industrial bonds.

C. Discount Rates and Tax Lives for Analysis

The preceding are the basis for assumptions used in the calculations that follow Thus, in the following analysis, cogeneration will be assumed to be owned by the forest products industry, and the discount rate employed will be 14 percent. For condensing power, utility ownership and a discount rate of 12 percent will be assumed. Power costs will be over the tax life of the project: 10 yr for a small (5 MW) cogeneration system assumed to be installed in a sawmill, 13 yr for a large (25 MW) cogeneration system assumed to be installed in a pulp mill, and 23 yr for condensing power plants assumed to be owned by utilities.

It is recognized that these cases do not present even project lives, however, costs not including capital (0 and M and Fuel) can be compared as well.

IV. THE COST OF POWER FROM WOOD FIRED FACILITIES

Cogeneration and condensing power costs now can be calculated for representative plant sizes using the data and formulas presented previously. These data are summarized in Table IX. Based on these data, pro forma statements can be created using Eq. (8-2) and the appropriate calculations can be made.

In order to construct such pro forma statements for pricing purposes, it is necessary to first calculate the required NATCF ($NATCF_r$) term using the following function:

$$CRF(I - ITC) = NATCF_r \qquad (8-7)$$

where CRF is the capital recovery factor (see Table IX for values used here, or any engineering economy text). From this information, required net income after taxes, earnings before taxes, and ultimately revenue can be developed by use of Eq. (8-2). Table X shows such a pro forma statement for the 5 MW cogeneration system.

Table X demonstrates the interplay of depreciation and profits in generating appropriate cash flows. For the first two years, profits may be negative. The system may take a loss. This is caused by the use of double-declining balance depreciation. Table X also may be used to calculate revenue requirements when operating cost categories are increasing at significantly different rates.

TABLE IX. BASE CASE ASSUMPTION SUMMARY
FOR POWER COST ESTIMATION

| Parameter | 5 MW | | 25 MW | |
	Cogeneration[a]	Condensing power	Cogeneration[b]	Condensing power
Gross capital investment ($\$x10^6$)	3.9	13.2	12.5	33.8
Investment tax credit ($\$x10^6$)	0.78	1.3	2.5	3.4
Net capital investment ($\$x10^6$)	3.12	11.9	10.0	30.4
Heat rate (MJ/kWh)	6.4	15.8	6.3	14.9
Representative load factor	0.80[a]	0.65	0.92[b]	0.65
Representative discount rate (percent)	14	12	14	12
Representative capital recovery factor 0.1917	0.1296	0.1710	0.1296	

[a]Sawmill.
[b]Pulp mill.

The most appropriate use of Table X, however, is when the
system is viewed as a "price taker"--when the revenue stream is
calculated by the function

$$P_M \times Q_P = R \tag{8-8}$$

where P_M is market price, Q_p the quantity of product pro-

TABLE X. PRO FORMA STATEMENT FOR 5 MW COGENERATION SYSTEM[a]

Cash flow stream	Year		
	1	5	10
Revenue	1280	1760	2370
Labor cost	80		
Supplies and other	10		
Fuel cost	290	920	1350
Maintenance	200		
Local taxes and insurance	100		
Depreciation	780	320	110
Earnings before taxes	(180)	520	910
Net income after taxes	(180)	280	490
Depreciation	780	320	110
Net after tax cash flow	600	600	600
Cost/kW (current dollars)	40 mills	50 mills	70 mills

[a]Values in dollars x 10^3.

duced, and YR and R the revenue in that year. In such a case the NATCF would not be calculated from Eq. (8-7) but would be calculated by Eq. (8-2) directly.

If, for example, a utility offered the cogenerator in Table X $0.055/kWh, then revenues would be $1,750,000, earnings before taxes would be $290,000. Net income after taxes would be $160,000, and NATCF would be $940,000. NATCF would climb to $1.04 x 10^6 in year 5, assuming an annual power price increase of 8 percent. Thus, NATCF is increasing at an annual rate of 2 percent. Under such circumstances, the NPV = $2.14 x 10^6 (DR = 14 percent) at the end of a 10 yr life.

If the market place for the power were $0.02/kWh, then the first year NATCF is only $20,000. Because profits are never

TABLE XI.   FIRST-YEAR ELECTRICITY COST FOR SUMMARY TABLE

	5 MW		25 MW	
	Cogeneration (TL = 10 yrs)	Condensing (TL = 23 yrs)	Cogeneration (TL = 13 yrs)	Condensing (TL = 23 yrs)
Capital charges ($\$x10^3$)	600	1710	1710	4370
O and M charges ($\$x10^3$)	390	1050	1030	3280
Fuel charges ($\$x10^3$)	290	680 [a]	2610 [b]	3470 [c]
Total ($\$x10^3$)	1280	3600	5350	11,120
First-year cost/kWh in 1980 dollars (to nearest $\$0.01$)[d]	0.04	0.12	0.03	0.08

[a]Fuel cost in $\$1.50/GJ$; heat rate is 16 MJ/kWh.
[b]Calculated by Eq. (8-4).
[c]Fuel cost is $\$1.80/GJ$; heat rate is 15 MJ/kWh.
[d]Assumes capturing every available tax credit.

generated and taxes never incurred, the annual rate of increase in NATCF is 8 percent. The NPV is $-\$3.01 \times 10^6$ (DR = 14 percent). The utility offered price is insufficient to warrant the investment.

Based on Table X it can be shown that a 1980 price of $0.04/kWh and annual price increase of approximately 7.5 percent is necessary to support the 5 MW cogeneration plant posited. As a practical matter the values in Table XI should be escalated by 25-30 percent for any plant now under study or design.

Table XI has been constructed showing first-year costs of power from cogeneration and condensing power modes. Condensing power costs are two to three times those associated with cogeneration. It should be cautioned, however, that these are generic mills whose fuel costs are at the base of the marginal cost curve shown in Fig. 8-1.

Table XI has been constructed using

$$P_R = [CRF(I - ITC) + C_{0,M} + C_F]/0 \qquad (8-9)$$

where P_R is the price required, $C_{0,M}$ the operating and maintenance cost (not including depreciation), C_F the fuel cost, and 0 the output. This holds only when no income taxes are required, a situation resulting from the use of double-declining balance depreciation, and capture its entire benefit.

If one assumes straight-line depreciation, no inflation, and a constant tax burden, Eq. (8-9) can be revised to

$$P_R = [CRF(I - ITC) + C_{0,M} + C_F + T]/0 \qquad (8-10)$$

where T is the taxes. Computation of the tax term is most easily made as a function of gross revenue. For utilities, a typical value would be approximately 10 percent of gross revenue as income tax (Internal Revenue Service, 1979). For the forest products industry, the comparable value would be approximately 5 percent of gross revenue rather than taxable income. For approximation purposes, then, the following expression holds:

$$P_R = [CRF(I - ITC) + C_{0,M} + C_F](1 - TR)/0 \qquad (8-11)$$

where TR is the tax rate.

Equations (8-9)-(8-11) apply only when stable depreciation, taxes, revenues, and inflation are assumed. These equations' utility is in providing a relative ranking of systems and in testing the sensitivity of systems to changes in fuel price, discount rate, or other terms.

Because the marginal cost curve for wood fuel has a step slope, Fig. 8-2 has been constructed to illustrate the influence of fuel price on power cost. Figure 8-2 is useful in summarizing technical influences affecting power price. The fact that cogeneration involves sequential steam and power production results in lower capital and fixed operating and maintenance costs. Thus, the intercepts on the vertical axis (the point where fuel values are 0) are lower for cogeneration systems, and at the same time as for condensing power systems, the slopes of the price curves are lower. Cogeneration is more fuel efficient and thus less affected by rising fuel price than condensing power. A 5 MW cogeneration system can pay approximately $6.00/GJ for fuel and compete with a 25 MW system paying about $1.00/GJ.

An alternative to this approach is to use the levelized cost model employed by utilities and described by Leung and Durning (1977). Two sets of fixed capital charge (FCC) values are required. For utilities, FCC values are 20 percent assuming a money cost of 12 percent. For the forest products industry, FCC values are 22 percent assuming a money cost of 14 percent. For simplicity, the cost of wood fuel is escalated at 7-1/2 percent, along with inflation. Table XII shows the levelized costs of power for 25 MW projects assuming a 20 year life.

Values in Table XII provide probably the best representation of power costs over the life of a project and demonstrate the relative economic positions of varius power generation options. These values are more realistic than first year costs, although they overstate first year power costs somewhat.

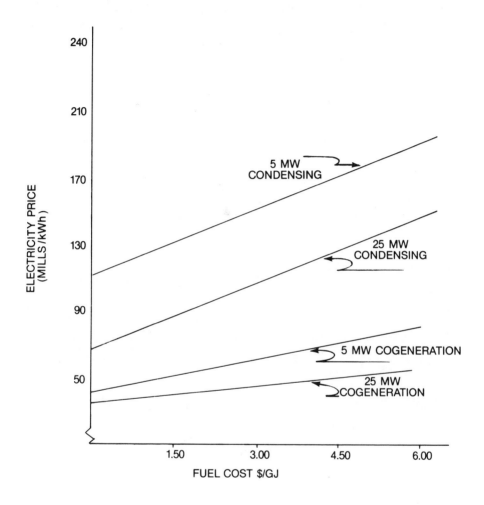

FIG. 8-2. The sensitivity of electricity of price to wood fuel cost and generating technology. Note that increased scale reduces price in both technologies examined. Note also that cogeneration typically produces a lower cost fuel than condensing power. Finally, the sensitivity of cogenerated power to wood fuel price is lower than the sensitivity of condensing power to wood fuel price.

TABLE XII.   LEVELIZED COSTS OF POWER FROM 25 MW COGENERATION
             AND CONDENSING POWER PLANT PROJECTS

	Cogeneration Project[a]	Condensing Power Project[b]
Capital charges ($x10^6$)	2.20	6.08
O and M charges ($x10^6$)	1.84	5.87
Fuel charges ($x10^6$)	4.75	6.18
Total ($x10^6$)	8.79	18.13
Output (kWh x 10^6)	200	142
Levelized Cost (mills/kWh)	44	130

[a]Forest industry ownership and operation at 92 percent load factor.   Incremental investment only.

[b]Utility ownership and operation at 65 percent load factor.

A determination of the competitiveness of wood-fired power can be made by comparing wood-produced power with that produced at a coal-fired condensing power plant. The latter costs approximately $0.05-0.06/kWh. Comparing this value with Table XII, one can see that 25 MW wood-fired cogeneration systems are highly competitive. The 5 MW cogeneration systems are marginally competitive. However, wood fueled condensing power systems generally are not competitive unless special conditions dictate their use. It is inefficient to produce electricity from wood in central station power plants if desirable fossil fuels can be used in systems capitalizing on thermodynamic gains from scale (e.g, higher pressures, reheat loops) and economies of scale (Tillman, 1980b).

V. CONCLUSIONS

Wood fuels are being used increasingly in the forest products industry and in the U.S. economy as a whole. They are used because wood is renewable, available on a year-round basis, and viewed by some as a cleaner fuel. Wood fuel is produced as a consequence of lumber, plywood, and pulp manufacture and can be used to raise steam and generate electricity required by processes for wood products manufacture.

This use of wood fuel, however, requires recognitiion of its inherent properties. It is both hygroscopic and anistropic. Relative to coal, it is modest in specific gravity and heating value, and moisture laden. It is, however, more reactive. Further, it contains less nitrogen, sulfur, and inert material.

Because wood fuels are derived from manufacturing processes applied to living materials, they are highly variable and somewhat difficult to handle. Their combustion involves numerous highly complex chemical processes including heating and drying, pyrolysis, gas phase combustion, and char oxidation. All such processes are influenced by firing method and fuel moisture content.

The process of wood combustion, employed to produce useful heat, has a thermal efficiency of 50-79 percent, depending on carbon conversion rates, moisture contents, and levels of excess air used. Typical efficiencies achieved are 65-68 percent for wet fuel. This contrasts with values of approximately 85 percent for coal. The process of wood combustion also results in airborne effluents such as particulates and oxides of nitrogen. While these are formed as a result of complex chemical phenomena, they are produced in modest quantities when compared to coal.

Combustion of wood, like coal, can be used to raise process steam or direct heat, although wood is not suited for large scale operations. Furthermore, wood can be used to produce electricity also--either in conjunction with process heat as cogeneration or as a primary product from condensing power plants. Typical heat rates are 6.3 MJ/kWh for cogeneration and 15 MJ/kWh for condensing power. These contrast with heat rates of 10 MJ/kWh for modern coal fired condensing power plants. The cost of power from wood fired systems is $0.04-0.06/kWh (1980 dollars) for cogeneration and >$0.10 (1980 dollars) for condensing power. These costs are highly sensitive to fuel price, and the marginal cost curves for wood fuels are steeply sloping. If wood fuels must be grown on silvicultural fuel farms, power prices would indeed be prohibitive. Currently, cogeneration is highly competitive with modern condensing power, however, wood fired condensing power requires an available fuel supply and a market for the steam being produced.

Wood fuels, then, are specialty fuels that can make significant contributions to the forest products industry in the raising of useful heat and electricity. Due to supply constraints, efficiency penalties, and costs, their use outside the forest products industry remains limited in extent.

REFERENCES

Alchian, A., and Allen, W. R. (1977). "Exchange and Production: Competition, Coordination, and Control," 2nd ed. Wadsworth, Belmont, California.

Bethel, J. S., et al. (1979). "Energy from Wood." A Report to the Office of Technology Assessment, Congress of the United States. College of Forest Resources, Univ. of Washington, Seattle, Washington.

Bierman, H., and Smidt, S. (1971). "The Capital Budgeting Decision," 3rd ed. Macmillan, New York.

Envirosphere (1980). "Preliminary Report: Feasibility of Cogeneration at the Georgetown Steam Plant." Envirosphere Co., Bellevue, Washington.

Feduccia, D. P., et al. (1979). Yields of unthinned loblolly pine plantations on cutover sites in the West Gulf region. U.S. Dept. of Agriculture, Forest Service, Research Paper SO-148, Southern Forest Experimental Station.

Fender, D. E. (1968). Short rotation, Proc. Symp. Planted Southern Pines, October 22-23, Cordele, Georgia.

Forest Industries Council (1980). Forest productivity report. National Forest Products Association, Washington, D.C.

Haley, C. W., and Schall, L. D. (1979). The Theory of Financial Decisions," 2nd ed. McGraw-Hill, New York.

Hass, J. E., Mitchell, E. J., and Stone, B. K. (1974). "Financing the Energy Industry." Ballinger Press, Cambridge, Massachusetts.

Internal Revenue Service (1977). "Tax Information on Depreciation." Department of the Treasury, Washington, D.C.

Jamison, R. L., Methven, N. E., and Shade, R. A. (1978). "Energy from Forest Biomass." National Association of Manufacturers, Washington, D.C.

Leung, P. and Durning, R.F. (1977). Power System Economics: on Selection of Engineering Alternatives. American Society of Mechanical Engineers, Joint Power Generation Conference, Long Beach, Ca., Sept. 18-21.

Schall, L., and Haley, C. W. (1977). "Introduction to Financial Management." McGraw-Hill, New York.

Standard and Poor's (1979). "Bond Guide, August, 1973." Standard and Poor's, New York.

Theusen, H. G., Fabrycky, W. J., and Theusen, G. J. (1977). "Engineering Economy," 5th ed. Prentice-Hall, Englewood Cliffs, New Jersey.

Tillman, D. A. (1980a). "A Financial Analysis of Silvicultural Fuel Farms on Marginal Lands." Ph.D. Thesis, College of Forest Resources, Univ. of Washington, Seattle.

Tillman, D. A. (1980b). Fuels from waste, in "Kirk-Othmer Encyclopedia of Chemical Technology," Vol. 2, 3rd ed. Wiley, New York.

Tillman, D. A., and Jamison, R. L. (1981). Cogeneration with wood fuels: A review, Fuel Processing Technol, (accepted for publication).

U.S. Census Bureau (1978). "Statistical Abstract of the United States 1978." U.S. Dept. of Commerce, Washington, D.C.

U.S. Census Bureau (1975). "Statistical Abstract of the United States 1975." U.S. Dept. of Commerce, Washington, D.C.

U.S. Forest Service (1979). "A Report to Congress on the Nation's Renewable Resources" (review draft). U.S. Dept. of Agriculture, Washington, D.C.

Value Line (1979). "Paper and Forest Products Industry." Value Line, New York.

Weyerhaeuser Co. (1979). Annual Report. Weyerhaeuser Co., Tacoma, Washington.

Weyerhaeuser Co. (1980). Annual Report. Weyerhaeuser Co., Tacoma, Washington.

INDEX